Quantitative Measures of Mathematical Knowledge

W0113211

The aim of this book is to explore measures of mathematics knowledge spanning K-16 grade levels. By focusing solely on mathematics content, such as knowledge of mathematical practices, knowledge of ratio and proportions, and knowledge of abstract algebra, this volume offers detailed discussions of specific instruments and tools meant for measuring student learning. Written for assessment scholars and students both in mathematics education and across educational contexts, this book presents innovative research and perspectives on quantitative measures, including their associated purpose statements and validity arguments.

Jonathan D. Bostic is an associate professor of mathematics education at Bowling Green State University, Bowling Green, OH, USA.

Erin E. Krupa is an assistant professor of mathematics education at North Carolina State University, Raleigh, NC, USA.

Jeffrey C. Shih is a professor of mathematics education at the University of Nevada, Las Vegas, USA.

Routledge Research in Education

This series aims to present the latest research from right across the field of education. It is not confined to any particular area or school of thought and seeks to provide coverage of a broad range of topics, theories and issues from around the world.

Recent titles in the series include:

For a complete list of titles in this series, please visit www.routledge.com/Routledge-Research-in-Education/book-series/SE0393

Quantitative Measures of Mathematical Knowledge

Researching Instruments
and Perspectives

Edited by Jonathan D. Bostic,
Erin E. Krupa, and Jeffrey C. Shih

LONDON AND NEW YORK

First published 2019 by Routledge

2 Park Square, Milton Park, Abingdon, Oxfordshire OX14 4RN

52 Vanderbilt Avenue, New York, NY 10017

Routledge is an imprint of the Taylor & Francis Group, an informa business

First issued in paperback 2020

Library of Congress Cataloguing-in-Publication Data
A catalog record for this book has been requested

ISBN: 978-1-138-59869-0 (hbk)
ISBN: 978-0-367-67075-7 (pbk)

Typeset in Sabon
by Apex CoVantage, LLC

Contents

Contributors

Amy Arneson's work focuses on investigating the efficacy of education policies and instructional programs on student outcomes and communicating research and evaluation findings to education leaders and practitioners. A central element of this work is the development of valid, reliable, and meaningful measures of student outcomes in the realms of academic achievement, social-emotional learning, and 21st-century skills. University of California, Berkeley. arnesonae@gmail.com

Jonathan D. Bostic's primary area of scholarship is exploring issues and trends within the context of quantitative assessment and evaluation in mathematics education. This area includes both development and uses of quantitative instruments for intended outcomes. Secondarily, he investigates ways to enhance instructional contexts to better support teaching and learning, especially learners' mathematical proficiency. This area includes work within inservice and preservice teacher education and professional development settings, as well as K-20 students' learning experiences and outcomes. Bowling Green State University. bosticj@bgsu.edu

Michele B. Carney is an associate professor of mathematics education at Boise State University. She specializes in mathematics teacher preparation and professional development. Her research focus is on how mathematics teacher education can be effectively scaled and measured. She has served as PI or co-PI on several large-scale mathematics professional development projects, including the NSF-funded Video Case Analysis of Student Thinking (VCAST). Boise State University. michelecarney@boisestate.edu

Laurie O. Cavey is a professor of mathematics education at Boise State University. She specializes in mathematical knowledge for teaching concepts related to algebra, functions, and mathematical modeling at the secondary levels. She enjoys investigating the ways that teachers think about students' mathematical ideas and how a focus on student thinking might transform the ways in which we educate teachers. She is currently serving as PI on the NSF-funded Video Case Analysis of

Student Thinking (VCAST) project and the NSF-funded project, Closing the Loop: From Student to STEM Teacher for Idaho Schools. Boise State University. lauriecavey@boisestate.edu

Jere Confrey's agenda is to see research applied to improve learning for all students through the use of learning trajectories and diagnostic assessment at scale. She hopes to see students' empowered by accessing their own data and see teachers make instruction more learner-centered. The way to accomplish this agenda is through deep collaborations among learning scientists, practitioners, and psychometricians. North Carolina State University. jconfre@ncsu.edu

Cassandra Hatfield's area of interest is in systems level change through leader and teacher professional development and coaching. These supports aim to impact teacher knowledge and implementation of research-based instructional practices. Southern Methodist University. chatfield@smu.edu

Michael D. Hicks' research interests include student thinking and reasoning about topics in abstract algebra. In particular, he is currently investigating the ways in which students might leverage connections between group theory and ring theory through reasoning by analogy. Texas State University. mh1653@txstate.edu

Heather Howell is a research scientist in the Student and Teacher Research Center at Educational Testing Service. Her research areas include the measurement and development of teacher content knowledge for teaching, with a focus in secondary mathematics teaching, and the study of teacher learning of educational practices such as discussion and argumentation. Educational Testing Service. hhowell@ets.org

Michael Kane conducts research in validity theory and practice, generalizability theory, and standard setting. Educational Testing Service. Mkane@ets.org

Leanne R. Ketterlin-Geller's research focuses on the development and validation of formative assessment systems in mathematics that provide instructionally relevant information to support students with diverse needs. She has engaged in validation work for many years, paying particular attention to the sources of validity evidence needed to draw formative inferences from classroom assessment data. She works nationally and internationally to support achievement and engagement in mathematics and STEM disciplines for students in kindergarten through grade 8. Southern Methodist University. lkgeller@smu.edu

Karl W. Kosko is an associate professor of mathematics education at Kent State University. His work focuses on how mathematical meaning is conveyed, focusing specifically on how children reason multiplicatively, the facilitation of mathematical argumentation, and how teachers engage in various representations of practice. Kent State University. kkosko1@kent.edu

Erin E. Krupa strives to make quality mathematics education more equitable to all students by researching the design, dissemination, and effectiveness of curricular materials and innovative professional development for mathematics educators and studying teachers' implementation of instructional materials. Erin's research pays close attention to the opportunity to learn students are provided within a classroom and how teachers can increase this index for all students, regardless of demographics. North Carolina State University. eekrupa@ncsu.edu

Matthew Lavery studies the development, validation, use, and improvement of educational assessments and psychological instruments. He investigates the analysis of data to support valid inferences and decisions that inform reflection and improve the outcomes of professional practice, programs, and policy in the areas of teaching, learning, and leadership. Bowling Green State University. mlavery@bgsu.edu

Patrick R. Lowenthal is an associate professor of educational technology at Boise State University, where he teaches master's and doctoral students in fully online graduate programs. He specializes in designing and developing online learning environments. His research focuses on how people communicate using emerging technologies, with a specific focus on issues of presence, identity, and community in online learning environments. Boise State University. patricklowenthal@boisestate.edu

Gabriel Matney researches authenticity in the learning and teaching of mathematics across world cultures with a purview on both classroom and professional development contexts. He seeks to develop instruments with sufficient validity evidence to be used in research on and about mathematics learning and teaching. Bowling Green State University. gmatney@bgsu.edu

Lindsey Perry's current research interests focus on investigating students' spatial and relational thinking abilities, developing mathematics assessments for young children, validating the interpretations of those assessments, and training educators on how to use data from assessments to make instructional decisions. Southern Methodist University. leperry@smu.edu

Jeffrey C. Shih is a professor at the University of Nevada, Las Vegas. Currently, he is interested in doctoral preparation in mathematics education and the development and use of quantitative measures. University of Nevada, Las Vegas. jeffrey.shih@unlv.edu

Elizabeth Stone is a research scientist in the Research and Development Division of Educational Testing Service (ETS). Her work since joining ETS in 1998 has included statistical analysis, coordination, and research for two major operational computerized adaptive testing programs; investigation of validity and fairness issues for students with disabilities and English learners on high-stakes and standards-based assessments; and the use of process data to enhance psychometric analysis. Educational Testing Service. estone@ets.org

Tatia Totorica is a clinical assistant professor for Boise State University's secondary STEM teacher preparation program, IDoTeach. She specializes in mathematics instruction, provides state-funded professional development for in-service secondary mathematics teachers, and serves on both the NSF-funded Video Case Analysis of Student Thinking (VCAST) and Closing the Loop: From Student to STEM Teacher for Idaho Schools grants. Boise State University. tatiatotorica@boisestate.edu

Emily Toutkoushian's research focus is on measurement in the context of classroom assessment, especially with regards to middle grades mathematics, validation, and technology. North Carolina State University. ektoutko@ncsu.edu

Diah Wihardini's research interests include item response modelling and test development, international assessments, teaching and learning, and policy analysis related to the aforementioned topics. Bina Nusantara University. dwihardini@binus.edu

Mark Wilson is a professor of education at the University of California, Berkeley, and also at the University of Melbourne. He teaches courses on measurement in the social sciences, multidimensional measurement, and applied statistics. His research focuses on the development of sound frameworks for measurement, new statistical models, instruments to measure new constructs, and philosophy of measurement. University of California Berkeley. MarkW@berkeley.edu

Acknowledgments

Reviewers

We wish to sincerely thank several reviewers who provided feedback on chapters:

> Jennifer Cribbs, Zandra de Araujo, Yanqing Ding, Rick Hudson, Marsha Ing, Cindy Jong, Eileen Murray, Samuel Otten, Robert Schoen, Nick Wasserman, Corey Webel, Ian Whitacre, and Annie Wilhelm.

National Science Foundation

This book is a product generated from a National Science Foundation funded conference held April 2–3, 2017 (Validity Evidence for Measurement in Mathematics Education; #1644314, 1644321). Any ideas expressed in this book by authors are those of the individuals and do not reflect the views of the National Science Foundation.

Personal

The editors wish to dedicate this book to their families.

> Jonathan—To Brynn, Matthew, and Josephine.
> Erin—To Marianne, Carolina, and Tyler.
> Jeffrey—To Meg, Abby, and Penelope

1 Validation in Mathematics Education

An Introduction to *Quantitative Measures of Mathematical Knowledge: Researching Instruments and Perspectives*

Erin E. Krupa, Jonathan D. Bostic, and Jeffrey C. Shih

Quantitative tools generate data resulting in analyzable numerical values that allow for statistical (inferential) analysis (Measurement in Mathematics Education Working Group, 2016). It is critical for a quantitative tool to measure the construct it intends to measure. Connections between a latent construct and how it is operationalized must be strong; otherwise, research foundations have potential to rest on spurious notions or unnecessary variance. For example, recently research indicated that free and reduced-price lunch (FRPL), which is often a measure of socioeconomic disadvantage, is actually not a valid operationalization for socioeconomic disadvantage relating to household income but may predict some type of educational disadvantage (Domina et al., 2018). It is important that quantitative measures have undergone a rigorous validation process for a similar reason. Without clear evidence to support the intended uses of a measure, claims made based on the measure are worthless, unsubstantiated, and potentially misinforming research. In her handbook chapter on research in mathematics education, Confrey states, "cumulatively the field loses credibility when researchers fail to meet the standards for rigor in research" (Confrey, 2017, p. 15). We argue the standards for the rigorous validation of measures' outcomes and careful attention to the uses of those measures in mathematics education need more attention by the research community, practitioner audiences, graduate education of future scholars in the field, and peer-reviewed journal articles.

The aim of this edited book and its companion, *Assessments in Mathematics Education Contexts: Theoretical Frameworks and New Directions* (Bostic, Krupa, & Shih, 2019), is to share descriptions of quantitative measures within mathematics education contexts, including purpose statements and validation arguments. A purpose statement is a concise narrative that describes the intent of the instrument and supports the use of the instrument in a particular context (Kane, 2001,

2012, 2016). In the *Standards for Educational and Psychological Measurement in Education* (*Standards*), the American Educational Research Association, American Psychological Association, and National Council on Measurement Education (2014) state, "A sound validity argument integrates various strands of evidence into a coherent account of the degree to which existing evidence and theory support the intended interpretation of test scores for specific uses" (p. 21). Chapters in this book weave together various validity evidence, from prior and current research efforts, to present sound validation arguments for interpretations of conclusions from specific quantitative measures.

A key focus in these chapters and the *Standards* is on interpretation of test scores for specific uses and not on the misinformed phrasing of "validity of the test". It is quite common among educational researchers and practitioners to refer to the validity of a test (Bostic, Krupa, Carney, & Shih, this volume). Further, there are reports from the larger field of educational scholarship that current conceptions of validity and validation are not widely used (Cizek, Rosenberg, & Koons, 2008; Shear & Zumbo, 2014; Wolming & Wikström, 2010). This book includes chapters addressing a wide range of quantitative measures within assessment and evaluation in mathematics education contexts and draws upon current conceptions of validity and validation in mathematics education. A central goal of this volume is to fill a needed gap in the field with a resource describing quantitative measures for mathematics education and their associated validity arguments. Thus, its chapters may be examples for others to follow, across diverse disciplinary backgrounds and career stages, as well as valuable descriptions of measures with validation arguments.

Importance of Validity

Measure quality strongly influences the quality of data collected and relatedly, findings of a research study (Gall, Gall, & Borg, 2007). Measures with a clearly defined purpose and supporting validity evidence are foundational to conducting high-quality quantitative work (Newcomer, 2009). A review of mathematics education research using quantitative measures presents a noticeable challenge: there are few syntheses of quantitative assessments for mathematics educators to employ and even fewer discussions of the validity evidence necessary to support the use of assessments in a particular context (Bostic et al., this volume). Quantitative measures lacking validity and reliability evidence may generate spurious results (Gall et al., 2007; Wilhelm, Gillespie, & Jones, 2018). Moreover, using a measure lacking validity and reliability evidence across multiple studies may lead to a research foundation built upon spurious results due to a flawed measure. It is advantageous for the scholarly community of mathematics educators and those exploring mathematics education phenomena to seriously consider their instrumentation and data collection as being grounded in a robust validation argument.

The heart of any methodology is the measure or instrument used to collect data (Newcomer, 2009). Scholars conducting quantitative research typically choose to either use a preexisting measure or develop a new one, depending on the purpose of the research. Informed decision making comes from having available data about quantitative measures. The *Standards* (American Educational Research Association et al., 2014) provide detailed guidelines regarding measurement validity and reliability. At a minimum, sufficient evidence for five sources must be shared related to validity: evidence from test content, evidence from response processes, evidence from internal structure, evidence for relationship to other variables, and evidence from consequences of testing (American Educational Research Association et al., 2014; Gall et al., 2007)—or a strong rationale for why evidence for one of those variables is unnecessary. For example, imagine a high school student taking a 10-item, constructed-response geometry problem-solving measure. If the measure uses items that have little or no content validity, then there is only a slim chance that the measure provides useful data about the student's geometry problem-solving performance. As another example, imagine that a mathematics content measure has low reliability; then the overall score may likely show greater evidence of random guessing than actual mathematics knowledge (American Educational Research Association, American Psychological Association, National Council on Measurement in Education, & Joint Committee on Standards for Educational and Psychological Testing (U.S.), 1999; American Educational Research Association et al., 2014; Hill & Shih, 2009; Shavelson, 1996). Additionally, if two students of equal ability but different ethnicities both take the same measure and score differently, then there is an issue of validity evidence related to other variables (in this case, ethnicity). Having any, or all, of these issues with validity evidence raises concerns about the interpretations of results from the measure. These examples also highlight the need to carefully and critically examine the validity and reliability evidence of measures used in research.

Unfortunately, "evidence of instrument validity and reliability is woefully lacking" (Ziebarth, Fonger, & Kratky, 2014, p. 115) in the literature. Validation studies of quantitative measures are noticeably absent from mathematics education journals too, which presents the challenge of determining whether an instrument is appropriate for a given study, much less whether its interpretations and results will be useful for analysis (Bostic et al., this volume; Hill & Shih, 2009). For instance, Hill and Shih (2009) reported that 8 of 47 studies published in the *Journal for Research in Mathematics Education* (JRME) in a 10-year time period provided any evidence related to validity and the majority provided only psychometric evidence. More recently, Bostic and colleagues (this volume) examined articles using quantitative research from two sources: JRME between 1970–2017 (*n* = 97) and a specific domain, namely early childhood research articles collected by the Development and Research in Early Math Education (DREME) Network between 2014–2018 (*n* = 24). They found most articles reporting student outcomes present no validity evidence (88%) related to the outcomes. Further, they report that of the 12% that did discuss validity evidence, the majority only mentioned test content and internal structure, and that when validity was discussed it was typically in reference to the measure, rather than the interpretation of the score.

Syntheses of measures for use in mathematics education can be found in the literature but are typically not intended as a comprehensive analysis. For example, Carney, Brendefur, Hughes, and Thiede (2015) conducted a brief review of self-report instructional practice survey scales applicable to mathematics education. The review was intended to provide a background on existing measures and their associated validity evidence in relation to a new measure under development. Boston, Bostic, Lesseig, and Sherman (2015) conducted a review of three widely known classroom observation protocols to assist mathematics educators in determining the appropriate tool for their particular research question and context. More recently, Bostic, Lesseig, Sherman, and Boston (in press) completed a synthesis of classroom observation protocols used in the last 25 years, which were appropriate for K-12 settings. It is important that this type of synthesis work continues and is encouraged by the field, which includes action by journal editors, university researchers, agencies that provide funding for research, and training for graduate students regarding how to conduct syntheses and metanalyses. In addition, we need to consider even more comprehensive approaches to ensure a breadth of topics and measures are examined.

In order for research in mathematics education to become more cumulative and to build on previous research, research instruments need to be widely shared, easily accessible, consistent in their intended use, and uniform in the interpretation of scoring. This book provides a much-needed resource for scholars making decisions about the use and interpretations of available quantitative measures, and it provides a comprehensive discussion of instruments used to measure students' mathematics content knowledge (Arneson, Wihardini, & Wilson; Confrey & Toutkoushian; Kosko; Perry & Ketterlin Geller, & Hatfield; Melhuish & Hicks, all this volume), mathematics instruction (Carney, Totorica, Cavey, & Lowenthal, this volume), and mathematics teachers' knowledge (Matney, Bostic, & Lavery, this volume).

Further, there are two chapters devoted to current trends in quantitative measurement of teachers' and students' competencies and future directions of that work (Bostic, Krupa, Carney, & Shih; Howell, Stone, & Kane, both this volume). Bostic and co-authors critically examine measures of students' knowledge and then raise particular opportunities for scholars to pursue in future research. Howell and colleagues identify challenging aspects of creating a validity argument and describe new approaches to assessing measures of teachers' competencies. They present three trends around teachers' competencies and describe important implications of these trends in relation to the construction of validity arguments.

Validity Arguments and Validation Evidence

This book and its companion, *Assessments in Mathematics Education Contexts: Theoretical Frameworks and New Directions* (Bostic, Krupa, & Shih, 2019), are intended to be accessible to a wide readership. These books are part of the dissemination efforts from a National Science Foundation funded conference, *Validity Evidence for Measurement in Mathematics Education (V-M²Ed)* (NSF #1644314), which brought together over 40 educational researchers working in mathematics education research with expertise in mathematics education, psychometrics, applied measurement, and special education to construct a shared understanding regarding validity within mathematics education contexts. Conversations centered on the perspective of validity articulated in the *Standards* (American Educational Research Association et al., 2014), including a strong focus on the meaningful operationalization of constructs into quantitative variables, and on Kane's (2006) model for articulating the purpose of an instrument with the proposed interpretation of test scores and necessary evidence to support the stated interpretations.

Five Sources From the Standards

Many of the chapters in this book link validity evidence to the five sources presented in the *Standards*: content, response processes, internal structure, relations to other variables, and consequences of testing (American Educational Research Association et al., 2014). Below we describe how authors in the chapters present evidence regarding the five sources. Note these are presented as examples to highlight some of the ways that this volume presents validity evidence connected with the five sources; however, the methods are neither exhaustive nor comprehensive. We encourage readers to examine each chapter for a more robust understanding of evidence regarding the five sources.

Test Content

Test content refers to "the relationship between the content of a test and the construct it is intended to measure" (American Educational Research Association et al., 2014, p. 14). Melhuish and Hicks (this volume) present a validity argument for a concept inventory for the Group Theory Concept Assessment (GTCA) that links validity evidence they collected to the five sources. In a very detailed and logical validity argument, they present five claims, one for each source of validity evidence, with subclaims for each that provides detailed explanation and results regarding the validity evidence they collected, and it supports their claims. One unique contribution of this paper is the use of a Delphi Study (Dalkey & Helmer, 1963) in the validity argument for evidence related to test content. Utilizing a Delphi Study, experts in group theory completed a series of rounds to determine the standard topics covered in an introductory group theory course. Unlike an expert panel, after each round, the experts received a summary of responses from the entire group and were given time to reflect and respond to the summary, which provided them with time to clarify their views and consider alternatives.

Response Processes

The second source, response processes, is evidence that connects how test takers *may* respond to a test item and how they *actually* respond to the item. In creating an assessment to measure teachers' knowledge of the *Standards for Mathematical Practice* (CCSSI, 2010), Matney and colleagues (this volume) gathered evidence based on the *Standards*. They conducted cognitive interviews with a sample of teachers who might complete the instrument, namely pre-service and in-service teachers as a source of response processes evidence. The cognitive interviews provided them with rich data regarding how the interviewees interpreted the items and offered ideas for revisions to the instrument.

Similarly, Kosko (this volume) presents a validation argument based on Kane (1992) and the *Standards* (American Educational Research Association et al., 2014) for a Multiplicative Reasoning Assessment (MRA) that he created based on atypical representations of multiplication and division. He also utilized cognitive interviews as a way to gather response process evidence. Additionally, he analyzed student written work to ensure students were responding in the intended ways to the MRA items, notably that they utilized one of four different strategy types as they solved items on the MRA.

Perry and colleagues (this volume) designed the Algebra Readiness Progress Monitoring (ARPM) assessment to measure middle school students' algebra readiness. They present an argument to align evidence based on response process to analyses conducted on data collected from students taking the APRM. Their analyses of response processes data included two studies: (1) a subject-matter expert review and (2) analyzing students' responses on flipped items. For the latter study, students responded to original items and then also flipped items, where the expressions were switched across the equal sign so the expressions on the left and right were switched from the original expression. The flipped item study was administered to provide evidence regarding a claim about student reasoning that was elicited from the APRM items. Together, these studies provide response processes evidence about the ARPM that support its interpretations from results.

Internal Structure

The third source presented in the *Standards* is evidence based on internal structure. Internal structure relates to "the degree to which the relationships among test items and test components conform to the construct on which the proposed test score interpretations are based" (American Educational Research Association et al., 2014, p. 16). This has important implications for the ways in which score interpretations are made. For example, it has implications for the manner in which a construct is being measured, psychometric fitting, and potential subscales of the test.

Arneson, Wihardini, and Wilson (this volume) utilize the Berkeley Evaluation and Assessment Research (BEAR; Wilson, 2005; Wilson & Sloane, 2000) assessment development framework to design a measure of students' college-ready statistical thinking along a developmental progression. The four building blocks they used in their design include construct map, item design, outcome space, and a measurement model (Wilson, 2005). In their conclusions, they link their validity evidence to the five sources. Specifically, they provide Wright Maps to claim response patterns conform to their hypothesized levels of the construct under study, which offers evidence related to internal structure. Additional evidence based on internal structure might include *differential item functioning*, to analyze how subgroups of test takers perform, or confirmatory factor analysis, to analyze if different components of the conceptual framework are in fact unique. These forms of validity evidence often come from psychometric results and may require collaboration with colleagues who have unique methodological training; to that end, we encourage collaborations among scholars from various backgrounds around a shared purpose, like assessment design.

Relations to Other Variables

The fourth source of validity evidence – evidence based on relations to other variables – connects scores and outcomes to other constructs, variables external to the test, and should "provide evidence about the degree to which these relationships are consistent with the construct underlying the proposed test score interpretations" (American Educational Research Association et al., 2014, p. 16). The *Standards* describe three facets of this evidence type: convergent and discriminant evidence, test-criterion relationships, and validity generalization. First, convergent and discriminant evidence analyzes the relationship between test scores and additional measures that are either intended to measure the same (convergent) or different (discriminant) constructs. For example, scores from an open-ended response test of student knowledge about fraction addition might be expected to relate closely to a measure of equipartitioning from a multiple-choice test. The relationship between the two measures, from assessments utilizing different methods, would provide convergent evidence. For discriminant evidence, it would be important to collect data from a construct that it is expected to relate less closely, such as shape sorting in the previous example. Second, a test-criterion relationship is the ability of the test scores to predict criterion performance, where the criterion is an outcome unique from the test. Third, validity generalization is "the degree to which validity evidence based on test-criterion relations can be generalized to a new situation without further study of validity in that new situation" (American Educational Research Association et al., 2014, p. 18). This evidence is in response to prior calls that researchers always had to perform new validation studies in their local contexts. However, validity generalizations harness the power of meta-analysis to argue there are circumstances where new validation studies do not have to be conducted and studies of the generalizability of validity evidence can be conducted.

It is important to determine if the scores produced by an assessment relate in predictable ways to other constructs that are known by the research community. Evidence for this source is often gathered after an instrument has been created and piloted (American Educational Research Association et al., 2014). Hence, it is important and should be explored in a timely manner.

Consequences of Testing

Finally, the fifth source – consequences of testing – explores the degree to which anticipated consequences from administering a test (and score outcomes) align with an intended purpose of the test. The *Standards* (American Educational Research Association et al., 2014) describe three specific consequences of testing: the interpretation and uses of test scores intended by test developers, claims made about test use that are not directly based on test score interpretations, and consequences that are unintended (pp. 19–20). Several of the chapter authors mentioned that collecting evidence regarding consequences of testing is an ongoing process. Melhuish and Hicks (this volume) present two claims to support their validity argument regarding the GTCA, which draw upon consequences evidence. Similar to relations to other variables evidence, consequences evidence is typically explored after an instrument's initial use (American Educational Research Association et al., 2014). This is often the case because it is uncertain what the actual consequences from test administration are until time has elapsed. Nonetheless, this sort of evidence should be explored as soon as possible because it is an important source.

Additional Validation Frameworks

Readers interested in newly developed validation frameworks should consult the companion to this book, *Assessment in Mathematics Education Contexts: Theoretical Frameworks and New Directions* (Bostic et al., 2019). In that book, readers learn about some new frameworks and ways they have been utilized in assessment development and use. In this book, we highlight two frameworks besides the *Standards* that chapter authors utilized. Kane advocates for an argument-based approach to validation that he refers to as an interpretation-use argument (IUA) and validity argument (VA) (Kane, 2012). The IUA includes a series of justifications and takes the form of a Toulmin (1958) model of reasoning with datum, claim, and warrants. The goal of the IUA is to justify the claims made regarding the interpretations and uses of scores. Then, the VA provides the evidence to support the claims and is "an evaluation of the plausibility of the IUA" (Kane, 2016, p. 69). Mathematics educators have used Kane's IUA and VA argument-based approach to validation to different degrees (e.g., Bell et al., 2012; Carney, Cavey, & Hughes, 2017). Two authors, Kosko (this volume) and Perry and colleagues (this volume) situate their validity argument in a "Kanian perspective" (Kane, 1992, 2012).

Pellegrino, DiBello, and Goldman (2016) detail a validity argument for assessments used in the classroom and for the instructional decisions resulting from assessment results. Their framework includes three components: cognitive validity, instructional validity, and inferential validity. Based on this framework, Confrey and Toutkoushian (this volume) share a validation study on the digital classroom assessment system, Math-Mapper 6–8. The digital system was built around learning trajectories for student reasoning on the measurement of circles. Research-based learning trajectories, clinical interviews, and design studies provide evidence of cognitive validity. Their argument for instructional validity revolves around how the Math-Mapper 6–8 was implemented and its alignment to the curriculum. Finally, for inferential validity they utilized statistical inference to analyze students' performance on tasks. Their overall validation study of the learning trajectory uses Rasch analysis and stepwise regression to provide a chain of reasoning for five main claims in their validation argument.

Construct Maps

Several authors utilized the BEAR (Wilson, 2005; Wilson & Sloane, 2000) assessment development framework (Arneson, Wihardini, & Wilson; Carney, Totorica, Cavey, & Lowenthal; Confrey & Toutkoushian, all this volume). Readers are encouraged to explore Wilson and Wilmont (2019) in the accompanying *Assessment* book, which synergizes the BEAR assessment system and the *Standards* to create a new framework for assessment developers. Wilson (2005) presents four steps in the assessment development process, the first of which is to create a construct map. There are few examples of developing construct maps in the literature. Therefore, Carney and colleagues (this volume) provide a much-needed example for open-ended assessments, specifically to measure teacher attentiveness. Their chapter provides a detailed account for the creation of a construct map and the qualitative ordering of item responses.

Conclusions

Research on validation work is woefully lacking in peer-reviewed outlets. Now is the time to collaborate across disciplinary boundaries to create more robust and rigorous measures in mathematics education contexts and for dissemination outlets to consider these important for publishing. Creating sound measures is not solely the work of mathematics education researchers; it should include collaborations with scholars in learning sciences, cognitive science, psychometrics, research assessment and evaluation, policy, special education, and other fields. Drawing on the expertise of diverse scholars will improve the measures that are being used by researchers, graduate students, and practitioners. Moreover, such collaborations have capabilities of better understanding phenomena and having broader impact across disciplines examining the same phenomena.

There is value in the knowledge that quantitative research can bring to the field in terms of generalizability to educational practice when appropriately conducted (American Statistical Association, 2007; Coalition for Evidence-based policy, 2003; Hill & Shih, 2009). The American Statistical Association's report (2007) on *Use of Statistics in Mathematics Education Research* states:

> If research in mathematics education is to provide an effective influence on practice, it must become more cumulative in nature. New research needs to build on existing research to produce a more coherent body of work. . . . Studies cannot be linked together well unless researchers are consistent in their use of interventions; observation and measurement tools; and techniques of data collection, data analysis, and reporting.
>
> (pp. 4–5)

It is an explicit aim that this book and its companion, *Assessments in Mathematics Education Contexts: Theoretical Frameworks and New Directions* (Bostic et al., 2019), can be used to build more cumulative knowledge of the validation of measures' outcomes and their uses in mathematics education. Through building this cumulative knowledge, there is substantial potential for fostering intellectual merit because there is a better answer to the question: How does one know that the results generated from a measure are validly drawn and connect to the intended construct? Such cumulative knowledge has potential for broader impact and generating robust study claims that can be leveraged in future scholarship with more validity evidence behind those conclusions. This is also a call to create new trainings and coursework for graduate students, or scholars unfamiliar with validation, to increase the appropriate use of measures and to assist those interested in creating new measures. Quite simply, there is no downside to collaboration across scholars engaging in assessment development and selection, or supporting graduate education in this manner.

We also remind readers that most validity arguments are the result of numerous studies. Even in this book, there is a range in the amount of evidence presented in each chapter. Some researchers report one detailed facet of their validation argument (e.g., Perry and colleagues solely present on response processes), whereas others present an argument that includes evidence from all five sources (e.g., Melhuish and Hicks draw on evidence from many studies to present evidence relating to five sources). It is the accumulation of evidence from various sources for claims that builds a coherent validity argument, not a single study or report. Additionally, it is amazingly difficult to provide sufficient validity evidence for all five sources related to an instrument in a 25-page manuscript that is submitted to a journal; when done, this can leave out details. Single studies when linked together can push forth a coherent validation argument and have potential to build this cumulative knowledge for scholarship in mathematics education. Validation deserves more recognition in peer-reviewed journals across disciplines connected with mathematics education; this is the type of research that moves the field toward more rigorous and cumulative research. As such, we hope this book can be a reference for others to use measures with sufficient validity evidence or to create their own validity arguments for measures they have created.

References

American Educational Research Association, American Psychological Association, & National Council on Measurement in Education. (2014). *Standards for educational and psychological testing.* Washington, DC: American Educational Research Association.

American Educational Research Association, American Psychological Association, National Council on Measurement in Education, & Joint Committee on Standards for Educational and Psychological Testing (U.S.). (1999). *Standards for educational and psychological testing.* Washington, DC: American Educational Research Association.

American Statistical Association. (2007). *Using statistics effectively in mathematics education research.* Alexandria, VA: Author.

Bell, C. A., Gitomer, D. H., McCaffrey, D. F., Hamre, B. K., Pianta, R. C., & Qi, Y. (2012). An argument approach to observation protocol validity. *Educational Assessment, 17*(2–3), 62–87.

Bostic, J. D., Krupa, E. E., & Shih, J. (2019). *Assessment in mathematics education contexts: Theoretical frameworks and new directions.* New York, NY: Routledge.

Bostic, J., Lesseig, K., Sherman, M., & Boston, M. (in press). Classroom observation and mathematics education research. *Journal of Mathematics Teacher Education.*

Boston, M., Bostic, J., Lesseig, K., & Sherman, M. (2015). A comparison of mathematics classroom observation protocols. *Mathematics Teacher Educator, 3*(2), 154–175.

Carney, M., Brendefur, J. L., Hughes, G. R., & Thiede, K. (2015). Developing a mathematics instructional practice survey: Considerations and evidence. *Mathematics Teacher Educator, 4*(1), 1–26.

Carney, M., Cavey, L., & Hughes, G. (2017). Assessing teacher attentiveness to student mathematical thinking: Validity claims and evidence. *The Elementary School Journal, 118*(2), 281–309.

Cizek, G. J., Rosenberg, S. L., & Koons, H. H. (2008). Sources of validity evidence for educational and psychological tests. *Educational and Psychological Measurement, 68*(3), 397–412.

Coalition for evidence-based policy. (2003, December). *Identifying and implementing educational practices supported by rigorous evidence: A user friendly guide.* Washington, DC: Author.

Common Core State Standards Initiative. (2010). *Common Core State Standards for mathematics.* Washington, DC: Author. Retrieved from www.corestandards.org

Confrey, J. (2017). Research: To inform, deform, or reform? In J. Cai (Ed.), *Compendium for research in mathematics education research* (pp. 3–27). Reston, VA: NCTM.

Dalkey, N., & Helmer, O. (1963). An experimental application of the Delphi method to the use of experts. *Management Science, 9*(3), 458–467.

Domina, T., Pharris-Ciurej, N., Penner, A. M., Penner, E. K., Brummet, Q., Porter, S. R., & Sanabria, T. (2018). Is free and reduced-price lunch a valid measure of educational disadvantage? *Educational Researcher, 47*(9), 539–555.

Gall, M. D., Gall, J. P., & Borg, W. R. (2007). *An introduction to educational research* (8th ed.). New York: Allyn and Bacon.

Hill, H. C., & Shih, J. C. (2009). Examining the quality of statistical mathematics education research. *Journal for Research in Mathematics Education, 40*(3), 241–250.

Kane, M. T. (1992). An argument-based approach to validity. *Psychological Bulletin, 112*(3), 527.

Kane, M. T. (2001). Current concerns in validity theory. *Journal of Educational Measurement, 38,* 319–342.

Kane, M. T. (2006). Validation. In R. L. Brennan, National Council on Measurement in Education, & American Council on Education (Eds.), *Educational measurement.* Westport, CT: Praeger Publishers.

Kane, M. T. (2012). Validating score interpretations and uses. *Language Testing, 29*(1), 3–17.

Kane, M. T. (2016). Validation strategies: Delineating and validating proposed interpretations and uses of test scores. In S. Lane, M. Raymond, & T. M. Haladyna (Eds.), *Handbook of test development* (Vol. 2). New York, NY: Routledge.

Measurement in Mathematics Education Working Group. (2016). *Validity evidence for measurement in mathematics education.* Retrieved from http://measuresinmathed.org

Newcomer, K. (2009). Basics of design for evaluation of cohesion policy interventions. In K. Olejniczak, M. Kozak, & S. Bienias (Eds.), *Evaluating the effects of regional interventions: A look beyond current structural funds' practice* (pp. 155–170). Warsaw, Poland: Ministry of Regional Development.

Pellegrino, J. W., DiBello, L. V., & Goldman, S. R. (2016). A framework for conceptualizing and evaluating the validity of instructionally relevant assessments. *Educational Psychologist, 51*(1), 59–81.

Shavelson, R. J. (1996). *Statistical reasoning for the behavioral sciences* (3rd ed.). Boston, MA: Pearson.

Shear, B. R., & Zumbo, B. D. (2014). What counts as evidence: A review of validity studies in educational and psychological measurement. In B. Zumbo & E. Chan (Eds.), *Validity and validation in social, behavioral, and health sciences* (pp. 91–111). Cham: Springer.

Toulmin, S. E. (1958). *The uses of argument.* London: Cambridge University Press.

Wilhelm, A. G., Gillespie Rouse, A., & Jones, F. (2018). Exploring differences in measurement and reporting of classroom observation inter-rater reliability. *Practical Assessment, Research & Evaluation, 23*(4).

Wilson, M. (2005). *Constructing measures: An item response modeling approach.* Mahwah, NJ: Erlbaum.

Wilson, M., & Sloane, K. (2000). From principles to practice: An embedded assessment system. *Applied Measurement in Education, 13*(2), 181–208.

Wilson, M., & Wilmont, D. (2019). Gathering validity evidence using the BEAR assessment system: A mathematics assessment perspective. In J. Bostic, E. Krupa, & J. Shih (Eds.), *Assessment in mathematics education contexts: theoretical frameworks and new directions.* New York, NY: Routledge.

Wolming, S., & Wikström, C. (2010). The concept of validity in theory and practice. *Assessment in Education: Principles, Policy & Practice, 17*(2), 117–132.

Ziebarth, S., Fonger, N., & Kratky, J. (2014). Instruments for studying the enacted mathematics curriculum. In D. Thompson & Z. Usiskin (Eds.), *Enacted mathematics curriculum: A conceptual framework and needs* (pp. 97–120). Charlotte, NC: Information Age Publishing.

2 The Form of Mathematics in Assessment Items

How Items Convey and Measure Multiplicative Reasoning Differently

Karl W. Kosko

Multiplication and division have been assessed in numerous formats by multiple entities over the past several decades. This includes large-scale, nationally representative samples through the National Assessment of Educational Progress (NAEP: Kloosterman, Mohr, & Walcott, 2016), the Trends in International Mathematics and Science Study (TIMSS: IES, 2017), the Programme for International Student Assessment (PISA: OECD, 2016), and others. Such large-scale assessments incorporate multiple choice and open-response item formats using word problem and symbolic representations of multiplicative contexts. Visual representations are also incorporated but are generally associated with a word problem context and/or symbolic expression. Furthermore, multiplication/ division items in large-scale assessments are constructed in reference to an overarching mathematics construct. A consequence of focusing, a priori, on multiple concepts within the overarching mathematical construct in designing such assessments is that they may be more likely to assess procedural knowledge of multiplication/division instead of conceptual knowledge and multiplicative reasoning (Kosko & Singh, 2018b).

Mathematics education researchers have also assessed multiplication and division incorporating variations in item format, including, but not limited to, symbolic representations, word problems, and visual models (Anghileri, 1989; Barmby, Bolden, Raine, & Thompson, 2011; Ghazali & McIntosh, 2004; Hodkowski et al., 2016; Nunes, Bryant, Evans, & Barros, 2015). Although many item formats used by these scholars mimic those used in large-scale assessments, additional interview-based tasks have also been used. For example, Anghileri (1989) used a 6 × 3 array with students to observe how children solved a task. The use of interview protocols in certain assessments (Anghileri, 1989; Mulligan & Mitchelmore, 1997; Smith & Smith, 2006) provides a means of assessing students' multiplicative reasoning, which large-scale assessment formats lack. However, interview protocols require time-intensive procedures for data collection, thereby inhibiting their use with larger samples.

The Multiplicative Reasoning Assessment (MRA) was designed to have the scalability of large-scale assessment designs while targeting the same form of reasoning assessed through interview protocol-based measures (Kosko & Singh, 2018a). To do this, items on the MRA include visual representations that elicit particular sets of actions (i.e., schemes) that are probabilistically associated with specific responses to said items. For example, given a rod that has a designated length of 3, a child is prompted to find the length of a juxtaposed rod, five times the length of the 3-rod (see the Type 3 item in Table 2.2, later in the chapter). In a typical assessment, use of such a visual representation includes accompanying symbolic expressions, an associated real-world context, or some combination of the two. The MRA breaches this norm, thereby presenting items that are non-canonical in format.[1]

The use of items deviating from generally accepted formats introduces questions of validity; namely, whether the new format of items assess what they are proposed to assess. This chapter discusses the content validity of MRA items primarily through evidence from students' response processes. In doing so, I present a portion of an overarching validity argument (Kane, 1992, 2012) with other portions of the validity argument presented elsewhere (Kosko, 2018a; Kosko, 2019; Kosko & Singh, 2018a). Thus, the primary purpose of this chapter is to gauge whether there is sufficient evidence of content validity for such non-canonical representations.

Theoretical Lens for Multiplicative Reasoning

What is multiplicative reasoning, and how does it interrelate with what is traditionally considered as multiplication and division? Davydov (1991) claimed "the original, basic operations underlying multiplication consist in transfer from a small unit to a larger one and in finding their relation" (p. 28), with all aspects of multiplication (and division) involving changing the system of units one is engaging. Boulet (1998) suggests that this definition, which focuses on transforming units of count, allows for a more cohesive mathematical definition that applies to various number systems (rational numbers, real numbers, etc.). Steffe (1991) later integrated Davydov's conception of unit into the radical constructivist paradigm. Initially, students engage in additive and counting schemes that precede multiplication (Kosko, 2018b; Steffe, 1994). These are followed by three multiplicative concepts. Hackenberg (2010) defined the three multiplicative concepts through descriptions of unit coordination, where a *unit* is a re-presentation of a counting act or acts (Ulrich, 2015). This chapter considers multiplicative reasoning primarily through Hackenberg's (2010) descriptions of multiplicative concepts alongside descriptions of emergent multiplicative schemes (Kosko, 2018b; Steffe, 1994).

Children engaging in *emergent multiplicative reasoning* (EMR) can coordinate one level of unit in activity. This generally involves variations of counting by 1s or repeated addition. A child operating at EMR solving 4 × 7 may separate counters into four groups with seven 1s in each group, and count each unit of 1 to construct the answer of 28. Children at EMR can often symbolically represent their actions with a multiplicative expression or equation (Davydov, 1991; Steffe, 1994) but do not engage in the operation of multiplication since they do not conceptualize the group of seven 1s as a unit in itself.

The multiplication operation is first observed when children engage in the *first multiplicative concept* (MC1). Children at MC1 can anticipate one level of unit to construct two levels of unit in activity (Hackenberg, 2010). Thus, a child solving 4 × 7 may anticipate the unit 4 as equivalent to four 1s and count the 4 seven times (4, 8, 12, . . ., 24, 28). By anticipating 4 as a unit in its own right, such children demonstrate what Davydov (1991) suggests as transferring from a smaller unit (1s) to a larger unit (4s) in their counting. Furthermore, such students can simultaneously consider a 4 as four 1s.

The *second multiplicative concept* (MC2) involves the coordination of three levels of units in activity, by anticipating two levels of units (Hackenberg, 2010; Kosko, 2018a). A child at MC2 asked to solve 4 × 21 may recognize that it can be rewritten as 4 × 3 × 7, and decide to solve it through repeated addition of 28 + 28 + 28, an additive version of 3 × (4 × 7). By contrast, a child operating at the *third multiplicative concept* (MC3) can anticipate three levels of units. Thus, a child at MC3 may also recognize 4 × 21 is equivalent to 4 × 3 × 7, but also that it is equivalent to 4 × (20 + 1), or even (4 × 3) × 7 = 12 × 7. Such a child is able to recognize both variations simultaneously and consider which is more efficient. A summary of these levels of unit coordination is provided in Table 2.1. Although the constructs represent a specific hierarchy with regard to unit coordination and multiplicative reasoning, the age or grade level in which specific multiplicative concepts are observed varies from child to child and classroom to classroom. Collecting data in May 2015 from 168 second and third grade students, Kosko and Singh (2018a) observed that at least 83.3% of participants operated at EMR by the end of second grade, while 54.3% of participants operated at EMR by the end of third grade. MRA data collected between 2016 and 2018, including 700 students, extends these estimates for the end of third (EMR = 52.9%, MC1 = 24.0%, MC2 = 9.6%, MC3 = 13.5%), fourth (EMR = 30.6%, MC1 = 15.3%, MC2 = 14.1%, MC3 = 40.0%), and fifth grade (EMR = 29.1%, MC1 = 16.8%, MC2 = 14.7%, MC3 = 39.3%). These percentages are not representative of a national sample, and also represent pooled data collected at varying points in each academic year. Their inclusion here is meant to emphasize that children's multiplicative reasoning varies across grade level and other contexts.

Table 2.1 Levels of Unit Coordination Characterizing Multiplicative Reasoning

Construct	Definition	Example Description	MRA Example Item
EMR	Constructs one level of units in activity.	Solves 4 × 7 by counting to 28 by 1s.	
MC1	Anticipates one level of units to construct two levels of units in activity.	Solves 4 × 7 by counting to 12 by 4s.	
MC2	Anticipates two levels of units to construct three levels of units in activity.	Solves 4 × 21 by recognizing there are three 4 × 7s, and uses repeated addition to add 28 + 28 + 28.	
MC3	Anticipates three levels of units to construct multiple levels of units.	Recognizes 4 × 21 can be solved either by adding 4 × 20 + 4 × 1 or by multiplying 3 × (4 × 7).	

Prior Assessments of Multiplicative Reasoning

A primary aspect of the meaning of multiplicative reasoning as defined by Davydov (1991) and expanded upon by Steffe (1994) and Hackenberg (2010), is the notion of inference by the child on the value of a unit, or units. Davydov (1991) suggested that "every question that results in multiplication is a problem in changing the system of units" (p. 30), but that many multiplicative tasks do not require this change. Furthermore, Davydov notes that using either a visual or symbolic model does not necessarily convey a multiplicative context. A child shown a 3 × 8 array may attend to the multiplicative nature by skip-counting three 8s, or they may attend to the additive nature by counting each discrete object by 1s. In recent years, several mathematics educators have attempted to construct measures of multiplicative reasoning. For example, Ghazali and McIntosh (2004) constructed a measure of Malaysian students' number sense in multiplication and division using 18 items including "word problems, mental computation, written computation and creating problems from a given written computation" (p. 98) with both pictorial and symbolic representations. Developing an assessment of what they describe as multiplicative thinking, Siemon and colleagues (Siemon & Breed, 2006; Siemon, Izard, Breed, & Virgona, 2006) included similar types of problems. For example, one task described by Siemon et al. (2006) asked students, if "a snail travels at 15 cm/minute, how far will it travel in 34 minutes?" (p. 117). Other tasks, described later by Siemon and Breed (2006), incorporate extended response formats embedded in real-world contexts. In their assessment items on multiplicative reasoning, Nunes, Bryant, Barros, and Sylva (2012) describe various examples including different representations and contexts. In one example, students are shown a two-length roll that has 8 sweets (candies) and are tasked with finding how many sweets are in a five-length roll. In a multiplicative reasoning assessment described by Barmby et al. (2011), a combination of symbolic representations and visual representations are included. Variation in the symbolic representations included recall facts such as "What is 5 times 2?" (p. 4) as well as relating multiplication expressions such as in the task "25 × 18 is more than 24 × 18. How much more?"

Each assessment briefly discussed in the prior paragraph includes a range of tasks involving multiplicative reasoning. For particular tasks, such as Siemon et al.'s (2006) snail task and Barmby et al.'s (2011) 5 × 2 task, students could respond with a rote procedure or memorized fact. However, many of the tasks involve a level of inference advocated by Davydov (1991). In the example provided by Nunes et al. (2012), the responding student is given that two rolls is 8, but must infer that one roll is 4 in order to deduce the value of five rolls. Consistent across each cited assessment is the large variance in the format of tasks provided within each assessment. Although it is clear that the tasks described by each group of researchers are situated in multiplication and division, few researchers provide an explicit test blueprint. Wilson (2010) notes that it is critical for test developers to identify descriptive components that "describe some aspect of the items . . . that are used to establish classes of items to populate the instrument" (p. 46). Framed in the context of construct validity, the test blueprint should specify how observable, measurable actions are indicative of descriptive components of the measured construct. As previously described, there is insufficient description in prior literature regarding the connection between item design and conceptualization of the construct. To clarify this point, I briefly discuss two different examples.

Siemon and Breed (2006) provide a detailed description of the learning progression they measure with their multiplicative reasoning assessment. This includes six observable actions: counting group items as 1s; modeling multiplication/division; abstracting multiplication/division; using derived and intuitive strategies for multiplication; using derived/ intuitive strategies for division; and extending/applying multiplication/ division with larger, more complex tasks. Siemon and Breed (2006) then describe how they used extended tasks (tasks with multiple components) to evaluate the different observable actions previously described using a specific assessment rubric. Unfortunately, the rubric design used relies on students explaining and illustrating their work in detail, which is rarely required of students on written assessments. Non-responses are not described with regard to this facet of item design. Contrasting the example provided by Siemon and Breed (2006) is that of Nunes et al. (2012). Items described by Nunes et al. (2012) prompt students for a single numeric response. The multiplicative portion of the assessment is based on Nunes et al.'s (2012) distinction between additive and multiplicative reasoning. For example, Nunes et al. (2012) assess additive reasoning with a length-based task for measuring how many centimeters long a ribbon is, and then assess multiplicative reasoning with a word problem for how many combinations of outfits Rebecca has when given 3 shirts and 3 pairs of shorts. The variation in forms of multiplicative reasoning is not addressed explicitly, and the differentiation in item design is focused only on additive versus multiplicative reasoning.

Both example assessments in the prior paragraph include items that provide evidence toward a validity argument (Nunes et al., 2012; Siemon & Breed, 2006). However, the descriptions of content validity provided at this stage of validation can be improved with regard to how conceptualization of the construct informs specific components of item design. This is not to suggest that either measure's outcomes are insufficient. Rather, I suggest that the construction of validity arguments, as advocated by Kane (2012) and others, should occur over several studies in order to provide adequate warrants for the claim of inference for a measure. Tzur and colleagues provide a useful example of this approach in their ongoing development of an assessment of various multiplicative schemes (Hodkowski et al., 2016; Johnson et al., 2018; Tzur et al., 2017). Rather than attempt to include all aspects of the validation process in a singular study, Tzur and colleagues have sought to construct their validity argument over a series of publications. Construction of the validity argument for the MRA has taken a similar approach.

In the next section, I provide an overview of the MRA. Like the aforementioned measures, the MRA has not constructed all aspects of a validity argument, and facets not examined may be identified as easily here as they have been for other measures involved in the ongoing validation process. Where the MRA distinguishes itself from other measures is that more work has been conducted on the development of the construct validity portion of the validity argument. I summarize this process in the next section.

The Multiplicative Reasoning Assessment

Over a 3-year period (2015–2018), the MRA was administered to a pooled sample of 868 students in second through fifth grade in a series of studies to collect evidence toward a validity argument. Describing the importance and criterion for constructing a validity argument for a measure, Kane (1992) suggested that

> because it is not possible to prove all of the assumptions in the interpretive argument, it is not possible to verify this interpretive argument in any absolute sense. The best that can be done is to show that the interpretive argument is highly plausible, given all available evidence.
>
> (p. 527)

As a measure is refined and improved over time, additional evidence accumulates for different facets of the validity argument. This chapter focuses on propositions associated with content validity, including:

- Items on the MRA elicit specific multiplicative schemes as instantiations of a unidimensional construct for multiplicative reasoning;
- Scores from the MRA associate with ability to complete and interpret conventional multiplication/division tasks.

These propositions relate both to the concept of multiplication and division, as well as the theoretical lens described as a basis for investigating multiplicative reasoning (i.e., the multiplicative concepts described by scheme theory). Prior administrations and versions of the MRA have presented evidence supporting these propositions (Kosko, 2018a, 2019; Kosko & Singh, 2018a). For many test developers, such evidence would indicate a need to focus on other aspects of the overarching validity argument. However, findings across several of these studies have revealed features about the nature of multiplicative reasoning (Kosko, 2018a, 2018b), informing revision and interpretation.

One approach to examining the content validity of an assessment is to collect parallel forms of evidence and to evaluate such data as applied to a different form (Kane, 1992; Wilson, 2010). While such an examination provides validity evidence for specific sets of items, it also provides evidence toward the content validity of the test blueprint and specific item design parameters. Such is the approach in this chapter. In order to properly frame such a study, I provide a brief overview of prior content validity efforts for the MRA as well as development of alternate forms in this process.

Versions of the MRA have been piloted with 868 students enrolled in grades 2 through 5 between May 2015 and October 2018. This includes the first pilot of items in May 2015 (n = 168), followed by pilots of smaller and larger samples in January and February 2016 (n = 27), May 2016 (n = 196), August and September 2016 (n = 41), September and October 2016 (n = 54), May 2017 (n = 210), February and March 2018 (n = 72), and October 2018 (n = 100). These various administrations of the MRA examined variations in item format and design that improved the item design blueprint for the assessment, with attention to the form of representations used in the MRA items. Additionally, students' written work on the assessments was digitally scanned and analyzed across the administrations, and students were interviewed regarding their multiplicative reasoning at key points in the validation process. Students discussed in the chapter are named with pseudonyms.

The MRA was initially piloted in 2015 with 168 second and third grade students (Kosko & Singh, 2018a). The initial version of the MRA included only length models of multiplication and division, and no symbolic representation of the multiplication/division operations.[2] By excluding other models (set, area) or problem types (word problems, symbolic expressions), the goal was to limit the "amount of unnecessary variance in item responses" (p. 12). Rather, there was a concern that including numerous item formats and representations could inadvertently measure other constructs not primarily related to multiplication or division. For example, past research has found that readability of word problems on NAEP and TIMSS assessments has differential effects based on students' backgrounds (Walkington, Clinton, & Shivraj, 2018). Therefore, creating items with as few confounding factors as possible was paramount.

Preliminary evidence of content validity was established through item analysis correlating students' response processes, as evidenced by their written work, with their correct/incorrect responses on particular items. Although classical item analysis suggested items had good discriminating power, the estimated Cronbach's alpha value of 0.79 did not meet the recommended threshold of 0.90 for cognitive assessments (Crocker & Algina, 2006). Further analysis of students' response processes indicated that many students who demonstrated appropriate strategies in their written work often miscounted by one or two skip-counts (Kosko & Singh, 2018a). To reduce the prevalence of response error, a reference line was included on each item (see Table 2.2 for example items including the reference line). Additional items were included to measure higher levels of reasoning (i.e., MC3).[3] The revised measure, hereinafter referred to as Form A, was administered to 196 fourth and fifth grade students in May 2016 and modeled using a Rasch approach (Kosko, 2019). As described by Bond and Fox (2015), Rasch modeling transforms raw, ordinal data into logit-based continuous measures linking item difficulty scores (delta statistics) probabilistically with student ability measures (theta statistics). Using students' written work and clinical interview data, Kosko (2019) found that students' theta scores aligned with the hypothesized levels for multiplicative reasoning. Additionally, students' written work has also provided evidence for content validity of Form A items (Kosko, 2018a).

During the 2016–2017 academic year, two additional forms were created: Forms B and C. Form C was designed to examine the application of the item design parameters to set and area representations, and was equated with Form A to examine whether the visual model used affects the psychometric properties. It is discussed in detail by Kosko (2018b). Form B was designed as a parallel form to Form A, using the same test blueprint and focusing on only length models and the same number of items per targeted scheme. It was also psychometrically equated with Form A using common anchor items. Following initial piloting with two classroom samples, Form B was administered to 132 third and fourth grade students. Rasch modeling of Form B estimated both sufficient item reliability (.97) and person reliability (.86), suggesting this version of the MRA had similar psychometric properties as Form A. With the exception of one item (Length_33; MNSQ = 1.38), all items had mean-square infit statistics inside of the accepted range (0.75 to 1.33), indicating unidimensionality (Bond & Fox, 2015). The one outlier item was examined following steps provided by Linacre (2010) and was found to be acceptable for inclusion in the measure. Form B was then equated with Form A using five anchor items (Length_02, Length_08, Length_10, Length_17, and Length_19). Range scores for particular levels were previously estimated with Form A (EMR is < –1.50; MC1 is –1.50 to –0.51; MC2 is –0.50 to 1.00; MC3 is > 1.00). Delta statistics for Form B suggest items typically align within hypothesized ranges. As noted by Kosko (2018b), some items designed to assess certain schemes may target other levels due to the magnitude of operands involved (Length_08, Length_25). A summary of item statistics of Form B, following equating, is provided in Table 2.2.

Table 2.2 Descriptive Statistics From Form B With Six Example Items

Item	Construct (Type)	Delta	SE	Infit MnSq	Outfit MnSq	Example Item
Length_02*	EMR	−1.25	0.24	1.00	0.77	
Length_20	(Type 1)	−2.18	0.25	0.95	2.81	= 1
Length_21		−2.18	0.25	0.83	0.79	=
Length_22	MC1	−1.04	0.25	0.78	0.54	= 5
Length_23	(Type 2)	−1.52	0.24	0.86	0.95	=
Length_24		−1.34	0.24	0.82	0.80	
Length_08*	MC1	0.25	0.27	0.65	0.45	= 3
Length_25	(Type 3)	0.65	0.28	1.17	0.99	=
Length_26		0.12	0.27	0.57	0.35	
Length_10*	MC2	−0.17	0.26	1.31	2.30	= 30
Length_11	(Type 4)	0.65	0.28	0.80	1.53	=
Length_27		0.19	0.27	1.06	1.69	
Length_28	MC2	0.82	0.29	1.27	1.36	= 12
Length_29	(Type 5)	0.99	0.29	1.07	0.84	=
Length_17*	MC3	1.64	0.32	0.68	0.30	= 24
Length_19*	(Type 6)	0.97	0.23	0.90	1.96	=
Length_31		1.54	0.31	0.84	0.70	
Length_32		2.07	0.34	0.83	1.26	
Length_33		5.14	0.57	1.38	9.90	

* Designates an anchor item for equating across forms.

In addition to various psychometric statistics, Table 2.2 also includes six example items used on Form B. The second column in Table 2.2 indicates the item type. Item type is a designation used in the initial validation of the assessment to designate the specific scheme an item is targeting (Kosko & Singh, 2018a). Across all items, students are provided a shaded given length and tasked with determining the length of the unshaded unknown length. The participating student must use the reference line and comparable lengths to coordinate units in accordance with the specific item type. Item Type 1 assess an emergent multiplicative scheme where a student must construct a composite unit by iterating units of 1 (coordinating one level of unit, in activity). For example, Length_20 provides the shaded length 1 and asks the student to find the unknown length ($1 \times 4 = 4$). Item Type 2 involves reversibility of the first item type, assessing whether a student can find a unit of 1 from a given composite length. Length_23 provides the given length 5 and tasks the participant to partition the 5 five times to find the unknown length's value ($5 \div 5 = 1$).

Item Type 3 assesses whether participating students will iterate a given composite length to find the unknown. For example, a student responding to Length_25 must anticipate one level of units by assuming a single length can be 3 and then iterate the given length 3 five times to find the unknown value of 15 ($3 \times 5 = 15$). Item Type 4 involves reversibility of Type 3, by tasking participants to partition the given composite into an unknown set of composite lengths. Length_27 illustrates this by juxtaposing a given length of 30 to the reference line with five evident partitions ($30 \div 5 = 6$). This involves anticipation of two levels of units, which is elicited visually through coordination between the given length and the reference line. Thus, a student completing this task must anticipate that 30 can be divided. This is distinct from a fair-share item which would allow for 30 to be divided into 1s and equally shared into five fair groups. By representing division in this way, it restricts direct operation of 1s units. Therefore, such a task engages students in "changing the system of units" (Davydov, 1991, p. 30).

Item Types 5 and 6 involve disembedding. Disembedding involves partitioning a subunit from a given whole and using the subunit to find another unit without destroying the original whole (Hackenberg, 2010; Kosko, 2018a). In the context of multiplication, disembedding is most visible in associativity or application of the distributive property. The distinction between Types 5 and 6 is that Type 5 involves finding and operating with a subunit as if it was a unit of 1, while Type 6 involves subunits that are composite numbers (Kosko, 2018a). Consider Length_28 as an example of Type 5. In order to solve the task correctly, a student must use the reference line to determine that 12 can be partitioned into four lengths of 3. Each of these lengths can subsequently partitioned into three lengths of 3. Simultaneously coordinating 12 as four lengths of three lengths of 1 involves coordinating three levels of units. The correct response of 8 can be determined in several ways (3 + 3 + 1 + 1; 6 + 2; 2 × 3 + 2; $2\frac{2}{3}$ × 3), but among the potential methods are viable strategies that involve disembedding to find units of 1. Contrast Length_28 with Type 6 item Length_17. A student completing this item must be able to partition 24 into three lengths of 8 and then multiply 8 twice (24 ÷ 3 × 2 or 24 × $\frac{2}{3}$). A student might attempt to solve the task by finding and operating on units of 1, but they are unlikely to find the correct answer with such a strategy.

The description of item types provided in the preceding paragraphs includes examples of symbolic expressions that may represent the strategies some students use. In prior analysis of students' response processes on Form A, such expressions were observed of various participants. Yet symbolic expressions were not required of participants to provide. It follows that if the MRA assesses multiplicative reasoning as previously described, a child who has successfully solved a certain item type should be able to demonstrate reasoning that aligns with the provided symbolic and/or visual representations provided. However, this is not to suggest that higher-level multiplicative reasoning must correspond with symbolic representation. Students operating at EMR can write multiplication and division expressions, and even perform two-column multiplication by rote (Kosko, 2019) but do so without considering multiplication by Davydov's (1991) definition. Thus, students' responses to tasks incorporating symbolic representations of the form $a × b = c$ or $c ÷ b = a$ have the potential to serve as miscues for their multiplicative reasoning (Kosko & Singh, 2018b). In the pages that follow, data from Form B of student response processes is examined to determine if the visual models incorporated in the MRA elicit the multiplicative reasoning that is targeted or if they elicit miscues themselves.

Evidence of Content Adequacy Validity Through Response Processes

The *Standards for Educational and Psychological Testing* (American Educational Research Association, American Psychological Association, & National Council on Measurement in Education, 2014) state that test content "refers to the themes, wording, and format of the items, tasks, or questions on a test" (p. 14). Evidence toward content validity can come from "logical or empirical analyses of the adequacy with which the test content represents the content domain and of the relevance of the content domain to the proposed interpretation of test scores" (p. 14). An additional source of evidence stems from evaluations by expert judgments from authorities on the content domain. The latter type of evidence, *content judged validity*, is the most typical reported in validation arguments of mathematics content measures. The former type of evidence, *content adequacy validity*, is reported less often. In the next section, I use evidence from students' response processes as an indicator of content adequacy. If items on the MRA assess multiplicative reasoning, then students should engage in multiplicative actions corresponding to both their demonstrated multiplicative reasoning and aspects of the domain that specific items target.

Response Processes From Written Work

Analysis of students' written work on Form B of the MRA focused on whether students provided evidence of one of four types of strategies: manipulating counts of 1s, doubling/halving, manipulating composite counts, and disembedding. Examples of evidence toward these strategies is illustrated in Table 2.3, with attention to four students' responses (Anibal, Oretha, Chantell, and Mily) to two items (Length_25 and Length_17). One student, Anibal, shows evidence of using doubling and halving strategies on both items by providing written expressions of this action. Variations of this strategy include partitioning one of the lengths in half, or iterating a second part of a given length. Another student, Oretha, provides evidence of manipulating counts of 1s on both items. Oretha's response to Length_25 illustrates a counting-on strategy (i.e., there are four additional partitions beyond the given length), whereas her response to Length_17 shows counting all partitions of the longest length. Across the 132-student sample, there were numerous variations of manipulating counts of 1s including, but not limited to, counting all, adding onto or down from, designating every length as 1, counting self-created partitions or those on the reference line, and adding all perceived units of 1 across both lengths. Both strategies of doubling/halving and manipulating counts of 1s are conceptually multiplicative without necessarily representing multiplication or division. Specifically, Oretha is coordinating two levels of units in a manner that does not demonstrate anticipating a composite (one level of units). By coordinating two levels of units, she is able to engage in a multiplicative context, but she does not engage in multiplication since she is constructing the first level of units in activity.

Variations of manipulating composite counts included skip-counting, partitioning a length into equivalent counts, and/or writing a multiplication/division expression. Two students, Chantell and Mily, demonstrated different variations of this strategy for Length_25. In each case, these students are demonstrating evidence of engaging with the items through multiplication schemes. Chantell provided clear evidence of coordinating two levels of units through her illustration of five groups of three 1s, in which the group partition lines are darker than the individual partitions for 1s. Mily provided similar evidence, but wrote a symbolic 3 instead of drawing individual partitions for 1s. Note that the more abstract example in Table 2.3 is provided by Mily, who has a higher theta score.

Table 2.3 Examples of Students' Written Work on Two Items

Student Θ score	Length_25 [δ = 0.65]	Length_17 [δ = 1.64]
Anibal Θ = −4.82	$3 \times 2 = 6$ $6 \div 2 = 3$ = 3 = 6	$24 \div 2 = 12$ = 24 = 12
Oretha Θ = −3.51	= 3 = 4	= 24 = 6
Chantell Θ = −0.90	= 3 = .15	= 24 48 = 48
Mily Θ = 4.55	= 3 = .15 $3 \times 5 = 15$	= 24 $6 \times 4 > 24$

Evidence of disembedding was diverse, with some students including written expressions, symbolically labeled partitions on the diagrams, or some combination thereof. Mily in Table 2.3 illustrates a combination of evidence showing, symbolically, that they divided the longer length 24 by 3. They then iterated this smaller unit to obtain the value of the shorter length. Although this student did not provide a symbolically written expression for their entire strategy, there is clear evidence that they iterated 8 twice following their division. Further, the relationship between the iteration of 8 and division/partition of 24 is conveyed through three partitions on the diagram related to division by 3 in the equation.

The brief analysis of student strategies above provides examples of what different strategies may look like if students are engaging with more or less sophisticated multiplicative reasoning. The next step in analysis was to code evidence of students' strategies for each item on Form B of the MRA. Frequencies for each code, by item, are presented in Table 2.4. A surface level examination of the descriptive statistics suggests that items with lower delta statistics have higher frequencies of count of 1s strategies, and items with higher delta statistics have higher frequencies of disembedding strategies. This coincides with the design of specific items. For example, only seven items were observed to elicit disembedding strategies from students. Yet, these were the only items purposefully designed to elicit disembedding. Similarly, items with large frequencies of count by 1s strategies (Length_02, Length_20, Length_21) were designed either to assess identifying the length of a composite by iterating 1s, or to assess identifying a length of 1 in comparison to a composite length (Length_22, Length_23, Length_24). A similar item-by-item analysis indicates an increase in observed strategies when an item was designed to assess a scheme aligned with that strategy. This provides useful evidence toward the content adequacy of items and confirms prior analysis by Kosko and Singh (2018a). No item was designed to assess a scheme associated with doubling or halving, so the observation of these strategies suggests potential variance in responses.

Next, the variance in students' ($n = 132$) theta scores was examined regarding the specific strategies used on individual items. It was hypothesized that students demonstrating a strategy corresponding with a targeted scheme would have higher theta scores than students using a different strategy. A Kruskal-Wallis one-way analysis of variance was used, per item, to examine whether theta scores differed by observed strategy (Siegel & Castellan, 1988). Given that there were 19 comparisons, a Simes adjustment to p-values was used to avoid a Type I error (false positive), and to hedge against the Type II error (false negative), to which the Bonferroni correction is prone (Simes, 1986). There was a statistically significant difference for each item, suggesting that the distribution of theta scores varied significantly by observed strategy. Some items demonstrated outlier

Table 2.4 Evidence of Strategy Use Across Items on Form B

Item	Delta	Counts of 1s	Doubling/ Halving	Composite Counts	Disembedding	Kruskal-Wallis
Length_02*	−1.25	88.2% $\bar{\Theta} = 0.63$	9.8% $\bar{\Theta} = -4.38$	2.0% $\bar{\Theta} = -2.10$	–	.003
Length_08*	0.25	52.0% $\bar{\Theta} = -1.71$	12.0% $\bar{\Theta} = -2.73$	36.0% $\bar{\Theta} = 2.13$	–	< .001
Length_10*	−0.17	42.5% $\bar{\Theta} = -1.46$	12.5% $\bar{\Theta} = -4.38$	45.0% $\bar{\Theta} = 1.91$	–	< .001
Length_11	0.65	45.7% $\bar{\Theta} = -1.86$	15.2% $\bar{\Theta} = -3.51$	39.1% $\bar{\Theta} = 2.10$	–	< .001
Length_17*	1.64	42.9% $\bar{\Theta} = -1.53$	19.0% $\bar{\Theta} = -3.03$	14.3% $\bar{\Theta} = -1.95$	23.8% $\bar{\Theta} = 3.36$	< .001
Length_19*	0.97	53.7% $\bar{\Theta} = -1.83$	9.8% $\bar{\Theta} = -4.38$	7.3% $\bar{\Theta} = -0.68$	29.3% $\bar{\Theta} = 3.09$	< .001
Length_20	−2.18	90.7% $\bar{\Theta} = -0.16$	9.3% $\bar{\Theta} = -4.49$	–	–	< .001
Length_21	−2.18	90.4% $\bar{\Theta} = -0.33$	9.6% $\bar{\Theta} = -4.30$	–	–	< .001
Length_22	−1.04	81.1% $\bar{\Theta} = -0.14$	13.2% $\bar{\Theta} = -2.97$	5.7% $\bar{\Theta} = -0.06$	–	.040
Length_23	−1.52	82.4% $\bar{\Theta} = -0.12$	15.7% $\bar{\Theta} = -3.76$	2.0% $\bar{\Theta} = 0.97$	–	.001
Length_24	−1.34	82.4% $\bar{\Theta} = -0.08$	13.7% $\bar{\Theta} = -4.06$	3.9% $\bar{\Theta} = 0.20$	–	.001
Length_25	0.65	55.7% $\bar{\Theta} = -1.94$	6.6% $\bar{\Theta} = -3.35$	37.7% $\bar{\Theta} = 1.67$	–	< .001
Length_26	0.12	56.9% $\bar{\Theta} = -1.77$	7.8% $\bar{\Theta} = -4.49$	35.3% $\bar{\Theta} = 1.79$	–	< .001
Length_27	0.19	36.0% $\bar{\Theta} = -1.63$	16.0% $\bar{\Theta} = -3.24$	48.0% $\bar{\Theta} = 1.03$	–	< .001
Length_28	0.82	51.1% $\bar{\Theta} = -1.67$	17.0% $\bar{\Theta} = -2.90$	8.5% $\bar{\Theta} = -1.31$	23.4% $\bar{\Theta} = 3.35$	< .001
Length_29	0.99	45.2% $\bar{\Theta} = -1.50$	19.0% $\bar{\Theta} = -3.02$	7.1% $\bar{\Theta} = -1.83$	28.6% $\bar{\Theta} = 3.07$	< .001
Length_31	1.54	44.7% $\bar{\Theta} = -1.76$	15.8% $\bar{\Theta} = -3.43$	10.5% $\bar{\Theta} = -2.35$	28.9% $\bar{\Theta} = 3.32$	< .001
Length_32	2.07	43.9% $\bar{\Theta} = -1.46$	17.1% $\bar{\Theta} = -3.71$	7.3% $\bar{\Theta} = -1.80$	31.7% $\bar{\Theta} = 3.20$	< .001
Length_33	5.14	25.1% $\bar{\Theta} = -1.51$	21.6% $\bar{\Theta} = -3.21$	10.8% $\bar{\Theta} = -1.63$	32.4% $\bar{\Theta} = 2.87$	< .001

* Designates an anchor item in both forms.

Note: All p-values for Kruskal-Wallis were adjusted using Simes's procedure.

strategies from fewer than five participants, indicating a trend counter to the design of the items (Length_22, Length_23, Length_24). However, the overwhelming trend suggests that students with higher scores tended to use strategies that correspond with targeted schemes on individual items.

Response Processes From Cognitive Interviews

In order to better understand how students with specific theta scores reasoned multiplicatively, a subsample of seven fourth grade students engaged in one-on-one interviews in March 2018 lasting approximately 5–10 minutes. A goal of the interviews was to assess the multiplicative reasoning of students with a range of theta scores. It was reasonable to target fourth grade students midway through the academic year, as this is a point in the grades 2–5 trajectory where they have had exposure to various multiplication/division concepts, but many may not have solidified their conceptual understanding yet. More specifically, this population was targeted for interviews given the potential variance in theta scores the sample may have. For sake of space, I report on interviews with three students: Michael (Θ = 1.30), Susan (Θ = –0.26), and Theo (Θ = –1.26). Each student completed Form B two days prior to interviews. Each cognitive interview involved asking students to solve, one at a time, eight different tasks selected from Form C (four set items and four area items). Form C items used different multiplicative models (set and area instead of length), but it was also equated to Forms A and B for comparison (Kosko, 2018b). Thus, using these items allowed for assessing whether students with particular ability measures (theta statistics) demonstrated multiplicative reasoning on different tasks from the assessment. For sake of space and simplicity, I discuss analysis of responses to four items (see Table 2.5). Students were asked to find the answer to each task, to explain how they solved the task, and to provide a multiplication or division sentence that represented the task.

Theo

Theo's theta score (Θ = –1.26) suggests he is operating at MC1 (–1.50 to –0.50), which indicates he is able to skip-count and begin to make meaning with basic multiplication/division. Thus, when responding to Set_03, Theo quickly stated "18, um 6 equals 12, and then plus 6," describing not only his answer but how he arrived at 18. He then proceeded to work on Set_06 by counting each circle by 5s. After reaching the end of the first row, Theo muttered "20" and then, pointing to the first two rows, "so this is 40 . . . 40 three times." At this point, Theo paused and wrote down 40 three times in a two-column repeated addition algorithm. Although Theo was able to reach the correct answer of 120, he did not provide a multiplicative expression to represent his strategy. This suggests that Theo was able to use skip-counting to construct units for repeated addition but did not relate this second level of units multiplicatively (as indicated by his repeated addition algorithm).

Next, Theo began his work on Area_06 by counting "10, 20, 30, 40, 50, 60, 70, and then there are 2 left (pointing to each row)." I asked him what

Table 2.5 Four Tasks Used in Cognitive Interviews

Item Designation	Delta	Item			*Example Equation(s)*
Set_03	–1.04		=	6	$6 \times 3 = 18$
			=	_____	
Set_06	–0.06		=	5	$5 \times 6 \times 4 = 120$
			=	_____	
Area_04	1.39		=	**36**	$(36 \div 4) \times 9 = 81$ or $\left(\sqrt{36 \div 4} \times 3\right)^2$
			=	_____	
Area_06	0.23		=	72	$72 \div 9 = 8$
			=	_____	

he was counting by 10s, and he pointed to each row on the shaded figure. He then applied the "2 left" to the unshaded row and assigned the square unit the value of 2. Through repeated addition, Theo found an answer of 18 for the unknown. Recalling that a conceptual understanding of division is characteristic of MC2 level activity, Theo's actions are not unsurprising. Rather, Theo demonstrated his ability to skip-count and do repeated addition with large numbers, but his scheme is not yet reversible. This was also evident in his work on Area_04, in which Theo determined that the larger unknown was "two wholes and a half" of the given quantity (it is actually 2.25 times as large), but did not relate this to the given quantity. Thus,

Theo's responses to the latter two tasks confirms he is unable to coordinate three levels of units consistently, and is operating at MC1.

Susan

Susan's theta score ($\Theta = -0.26$) suggests she is operating at MC2 (-0.50 to 1.00), which should indicate she is able to engage meaningfully in single-digit multiplication and division tasks as well as engage with more complex multiplicative tasks involving disembedding. When solving Set_03, Susan quickly responded that it was 18 "because 6 times 3 equals 18." Moving to Set_06, Susan counted all individual elements by 5s and arrived at a miscount of 130 (the correct answer is 120). When prompted to provide a multiplication or division sentence for the task, Susan used 5×24. She explained that there were 24 counters, using her pencil to indicate the 6 row and 4 column structure, and that since she counted by 5s it would be 5×24. Although Susan did not find the correct answer, her explanation shows engagement with constructing three levels of units (at least MC2). Specifically, she was able to identify all individual elements as having a value of 5, and she was able to quickly identify that there were 24 elements by anticipating the 6×4 multiplicative structure in the array. Although able to construct three levels of units, Susan did not demonstrate that she anticipated three levels, as she relied on skip-counting 24 5s. Confirmatory evidence of Susan operating at MC2 comes from her response to Area_06, where she was able to explain the division explicitly.

Susan's response to Area_04, an MC3 task, provides further evidence that she is operating at MC2 and is not yet able to anticipate three levels of units. After counting each square unit in the unknown figure, Susan stated that it was 45 "because I added this (nine squares) to 36." When pressed to explain further, Susan could not interpret the nine squares in the unknown figure with the four squares having a value of 36 in the given figure.

Michael

Michael ($\Theta = 1.30$) was assessed by the MRA as operating at MC3 (1.00 and up), suggesting he is able to meaningfully apply the associative and distributive properties to solve multiplication/division tasks with single and multiple digits. After quickly responding that the answer to Set_03 was 18 because it represented 6×3, he began to work on Set_06. Michael used the tip of his pencil to count the outside rows and columns, and then wrote out the traditional algorithm for 24×5. Asked to explain, Michael stated that "first I did 6 times 4, 24. Then I multiplied 24 times 5." Whereas Susan counted by 5s before providing the expression 24×5, Michael used the expression to calculate his response of 120. Thus, he anticipated the multiplicative structure of rows × columns × the value of each discrete unit (three levels of unit).

After quickly solving Area_06 and relating the figure to both 9 × 8 = 72 and 72 ÷ 9 = 8, Michael began work on Area_04. Michael responded that the answer was 9 (the correct response is 81, or nine 9s). When pressed to explain, Michael stated "because 2 times 2 equals 4, and 4 times 9 equals 36, or you could have divided 36 by 4." Michael clarified his strategy by indicating he was looking at the 2 × 2 structure of the given figure, and then dividing 36 by 4 to find the value of 9. Although Michael did not use his found unit of 9 to find the correct value of the unknown (81), his explanation demonstrates anticipation of three levels of units. His strategy could be written symbolically as 36 ÷ (2 × 2) = 9, and he later equated 36 ÷ (2 × 2) with 36 ÷ 4. However, Michael's explanations and work also demonstrate that there are varying degrees of sophistication even at a MC3 level of reasoning. Michael was able to solve certain MC3 items correctly, and others incorrectly, but anticipated three levels of units in his work on all MC3 tasks during his interview.

Summary of Evidence From Students' Response Processes

Describing models for content validity, Kane (2012) suggested that

> where a sample of some type of performance . . . is used to draw conclusions about level of skill in that kind of performance, a good case for the validity of the proposed interpretation can be made on rational grounds.
>
> (p. 5)

Students' response processes were examined in data from written work from tasks on Form B of the MRA and from cognitive interviews with tasks not from Form C. The rationale for including both types of evidence was to include two different types of "performance." The first type, coded strategies from written work, demonstrated that students' theta scores tend to correspond with the use of specific strategies on particular items. Thus, given an item targeting MC2 level reasoning, students who use strategies aligning with MC2 also have higher theta scores than students who use strategies that are considered as aligning with MC1 or EMR. This trend was found to be statistically significant across reasoning types and items. Thus, it provides evidence that the items are eliciting the level of multiplicative reasoning that they were designed to assess.

The second type of "performance" examined how predictive students' theta scores from Form B were with regard to their engagement with multiplicative tasks not included on Form B (see Table 2.5). Students' strategies and explanations corresponded with the level of multiplicative reasoning indicative of their theta scores. Theo (Θ = –1.26) was assessed as operating at MC1 and demonstrated skill in skip-counting and basic multiplication, but he was not able to relate these actions to division. Susan (Θ = –0.26) was assessed as operating at MC2 and demonstrated an understanding of basic multiplication and division. However, when she was tasked with coordinating three levels of units, she represented the appropriate multiplication only after engaging in skip-counting. By contrast, Michael (Θ = 1.30), assessed as operating at MC3, was able to anticipate and operate on the multiplicative expressions associated with tasks involving three levels of units. Michael incorrectly solved Area_04 while demonstrating a strategy representative of MC3 level thinking. It is important to note that other students interviewed with higher theta scores than Michael were able to extend the same reasoning Michael conveyed to multiply the 9 unit nine times for a correct answer. Thus, evidence from cognitive interviews suggests that theta scores from the MRA are indicative of different levels of reasoning and the use of different strategies on particular tasks.

Discussion

The purpose of this chapter was to discuss whether, and how, non-canonical visual representations provide a valid means of assessing multiplication and division. Although prior study has presented evidence for the content validity of the MRA (Kosko, 2018a, 2019; Kosko & Singh, 2018a), some question whether such non-canonical visualizations represent the construct of multiplicative reasoning. Kane (1992) suggested that validity evidence is most needed, and most effective, when a particular aspect of a validity argument "can be questioned because of existing evidence indicating that it may not be true, because of plausible alternative interpretations that deny the assumption, because of specific objections raised by critics, or simply because of a lack of supporting evidence" (p. 530). For the MRA, the "objections raised by critics" focus on whether or not the visual models used by the MRA actually assess multiplicative reasoning. In this chapter, I have presented additional supporting evidence to address these objections. Parallel forms of evidence (students' written work and cognitive interview data on different multiplicative tasks) from Form B data presented in this chapter, and evidence presented from Form A in prior study (Kosko, 2018a, 2019), provide several sources of evidence that suggest the MRA assesses the domain of multiplicative reasoning. Pending a plausible counterargument supported by data, it is highly probable that the MRA has sufficient content validity. This is not to indicate that the propositions regarding the content validity portion of the MRA's validity argument are infallible. Kane (1992, 2012) has repeatedly stated that no validity argument is ever certain. Furthermore, validity arguments are dynamic and ever-evolving. "As new information becomes available, the [validity] argument may expand to include new types of inferences" (Kane, 1992, p. 534).

The content validity portion of the MRA's validity argument has now been established through several studies and across multiple sources of data. Given the amount of evidence collected and the perceived need for it, there are lessons to be learned for others wishing to construct a measure with similarly unique items or those wishing to adapt items from the MRA. In the sections that follow, I provide some recommendations for other researchers in this regard.

Constructing a Validity Argument Over Multiple Studies

It is clear from Kane's (1992) recommendations for constructing validity arguments that such construction need not include all forms of validity in any particular study. However, it is common for those engaged in the validation process to face criticism at the review stage for not including all forms of validity evidence that are conceivable. I remind the reader that the need for a validity argument is to provide evidence for the plausibility of specific propositions made regarding a measure.

To help exemplify the need for validity evidence, I return to two other measures of multiplicative reasoning discussed earlier in the chapter (Nunes et al., 2012; Siemon & Breed, 2006). The assessments described by Nunes et al. (2012) and Siemon and Breed (2006) include their theoretical framework for item design and provide some explanation for this alignment. However, their descriptions do not have the level of detail as prior descriptions of the MRA (Kosko, 2018a, 2019; Kosko & Singh, 2018a). A simple explanation for this is that the items used by Nunes et al. (2012) and Siemon and Breed (2006) represented more canonical forms of multiplication and division. Because items similar to the form used in such assessments have been applied by researchers and test developers over the past several decades, the plausibility that such items have sufficient content validity is high. Therefore, there is less need for evidence to support the content validity of those forms of items unless a data-supported counterargument suggests otherwise.

Including the current chapter, the development and construction of the content validity argument for the MRA is conveyed across several studies (Kosko, 2018a, 2018b, 2019; Kosko & Singh, 2018a). As evidence for content validity emerged, findings from some studies suggested further clarification of the construct itself. For example, Kosko's (2018a) analysis of disembedding items clarified the distinction between disembedding at MC2 versus disembedding at MC3. Another example comes from Kosko (2018b), who found that the magnitude of the operands used in multiplicative tasks interacts with the level of reasoning students demonstrate.

Thus, students may need MC2 reasoning when engaging with operands larger than 5, even when solving tasks that appear as if they should elicit MC1 (Kosko, 2018b). Neither set of findings from these two studies counters evidence toward the construct validity of the assessment. Rather, they lend support for the MRA's ability to measure aspects of the construct previously underspecified. This suggests a significant implication for other researchers engaging in similar processes of test development. Specifically, even when focusing on heavily studied constructs (such as multiplication and division), evidence toward construct validity may inform specification of the construct itself.

Recalling Kane's (1992) statement that less plausible propositions in a validity argument require more evidence, certain propositions may require multiple studies to describe such evidence comprehensively. The construct validity propositions for the MRA provide a useful example, but I remind the reader that other measures of the same construct (i.e., Nunes et al., 2012; Siemon & Breed, 2006) have not provided a similar scope of evidence toward construct validity. Thus, the particular features of a measure, its item design, and in some cases the construct itself suggest the level of evidence needed for particular propositions for validity.

Implications for Other Assessments

The findings presented here provide support for the notion of using non-canonical visual representations to assess specific mathematical concepts. There are two primary benefits of incorporating such representations in assessments like the MRA. First, such items allow for quickly scoring student responses to items as correct/incorrect. Second, if designed properly, correct/incorrect scores on such items can provide accurate probabilistic indicators of whether students demonstrate schemes targeted by the item's visual design. Further, by reducing the elements associated by canonical representations, student responses to such non-canonical items are more reliable indicators of their reasoning, as indicated both in this chapter and by prior study (Kosko & Singh, 2018b).

Design of non-canonical items described in this chapter focused on multiplicative reasoning, but other concepts and topics could also be assessed using very similar design parameters. Key in the test blueprint for the MRA and the design of specific item types (see Table 2.2) is the inclusion of a given unit and unknown unit that must be inferred by the student (anticipatory schemes as described by Hackenberg, 2010). This design feature borrowed from Davydov's (1991) idea of multiplication as determining the relationship between two quantities "not directly, but with an intermediary" (p. 20). By providing a given unit and modeling the visual scaffolds to elicit particular inferences, specific unit coordination actions were able to be assessed. Consider Length_25 illustrated in Table 2.3. The item provides the given unit 3 and visually juxtaposes a

1 cm length assigned this value with a 5 cm length alongside a reference line that partitions it into five 1 cm segments. We know from students' response processes that participants demonstrating at least MC1 level reasoning were able to represent this either as 3 skip-counted five times, or as 3 × 5. We also know that these same design principles apply to both set and area models (Kosko, 2018b).

The same item design principles used to elicit multiplication with Length_25 can be applied to other concepts and topics. To illustrate this point, consider the various ways that Length_25 can be revised to assess other mathematical topics. If a researcher wished to examine students' conception of improper fractions, they might revise the item so that the given unit is $\frac{1}{3}$ (instead of 3), resulting in a correct solution of $\frac{5}{3}$ or $1\frac{2}{3}$. A variation focusing on decimals could change the given to 0.2, in which the correct solution would be 1. A different variation might tell students the value of the shorter length is x and larger length is y, and then ask students to write an equation conveying the relationship between the two. The point of this brief exercise is not to convey the adaptability of items from the MRA but to convey how a focus on relationship between quantities, and not simply their symbolic form, can be realized by adequate structure of specific representations. In attending to such relationships, it is essential for test developers and educational researchers to attend to recommendations, such as those by Wilson (2010), to specify the particular aspects of a construct (theoretically defined) that are to be observed with specific actions on the assessment.

Conclusion

Evidence from students' response processes from written work and observations in interviews provides support for the descriptive components of the test blueprint for the MRA. This in turn supports the hypothesized relationship between the design of item classes and the specific facets of the construct those classes signify (Wilson, 2010). These same descriptive components of the test blueprint have previously been examined for an alternative form of the MRA (Kosko, 2018a, 2019; Kosko & Singh, 2018a). Demonstration of evidence toward the construct validity of the MRA supports inferences not only of the specific items included on the assessment but also of the test blueprint itself. I suggest that the synthesis of such evidence provides strong support for the construct validity of the MRA. Therefore, there is strong evidence for the claim that the non-canonical item format used in the MRA allows for meaningful and valid assessment of multiplicative reasoning at varying levels of sophistication.

This chapter provides several lessons for other researchers. First, the construction of a validity argument, or even a portion of such a validity argument, may span several studies. Also, the empirical evidence and out-comes produced from such studies may be valued in terms of validity, but not the assessment itself (the assessment's validity argument is based on the empirical evidence on not on the assessment itself). Evidence toward an assessment's validity argument may be contingent upon the novelty of the item design, such as with the MRA, with regard to the nature of the construct being measured, or both. Second, those engaged in procuring evidence toward the content validity of an assessment should follow closely the recommendations of scholars like Wilson (2010) in developing their test blueprint. Such specification allows for refinement not only of the assess-ment but potentially the construct itself. A final lesson from this chapter focuses on how we conceptualize the constructs we seek to measure. Such conceptualizations relate directly to how we write items. In the case of multiplicative reasoning, my own conceptualization of multiplication and division did not assume a necessary familiarization with the Westernized symbols for multiplication/division or the word problems that we may use to contextualize multiplication. Rather, the MRA focuses on a mathemati-cal definition of multiplication and division provided by Davydov (1991) and others. For mathematics educators engaged in developing assessments, a similar focus on mathematical meaning is necessary in order to lay our personal assumptions about the construct bare. The essence of construct validity lies in attending to the fundamental mathematical meaning con-veyed by the items we write.

Notes

1. Describing MRA items as non-canonical is meant to suggest a deviation from the norm of multiplication/division items including symbolic expressions/equations, world problems, and/or accompanying "real-world" contexts.
2. Other visual models have since been incorporated in the most recent version of the MRA.
3. A small pilot involving two fourth grade classes compared response processes when the reference line was, and was not, included. Inclusion of the reference line appeared to reduce the amount of variance in student responses.

References

American Educational Research Association, American Psychological Associa-tion, & National Council on Measurement in Education. (2014). *Standards for educational and psychological testing.* Washington, DC: American Educational Research Association.

Anghileri, J. (1989). An investigation of young children's understanding of multi-plication. *Educational Studies in Mathematics, 20*(4), 367–385.

Barmby, P., Bolden, D., Raine, S., & Thompson, L. (2011). Assessing young chil-dren's understanding of multiplication. *Proceedings of the British Society for Research into Learning Mathematics, 31*(3), 1–6.

Bond, T., & Fox, C. M. (2015). *Applying the Rasch model: Fundamental mea-surement in the human sciences.* New York: Routledge.

Boulet, G. (1998). On the essence of multiplication. *For the Learning of Mathematics, 18*(3), 12–19.

Crocker, L., & Algina, J. (2006). *Introduction to classical and modern test theory.* Mason, OH: Wadsworth Thomson Learning.

Davydov, V. V. (1991). A psychological analysis of the operation of multiplication. In V. V. Davydov (Ed.), *Psychological abilities of primary school children in learning mathematics* (pp. 9–85). Reston, VA: National Council of Teachers of Mathematics.

Ghazali, M., & McIntosh, A. (2004). From doing to understanding: An assessment of Malaysian primary pupils' number sense with respect to multiplication and division. *Journal of Science and Mathematics Education in Southeast Asia, 27*(2), 92–111.

Hackenberg, A. J. (2010). Students' reasoning with reversible multiplicative relationships. *Cognition and Instruction, 28*(4), 383–432.

Hodkowski, N. M., Hornbein, P., Gardner, A., Johnson, H. L., Jorgensen, C., & Tzur, R. (2016). Designing a stage-sensitive written assessment of elementary students' scheme for multiplicative reasoning. In M. B. Wood, E. E. Turner, M. Civil, & J. Eli (Eds.), *Proceedings of the 38th annual meeting of the North American Chapter of the International Group for the Psychology of Mathematics Education* (pp. 1581–1587). Tucson, AZ: University of Arizona.

Institute of Education Sciences [IES]. (2017). *The condition of education 2017* (Research Report). Washington, DC: US Department of Education.

Johnson, H. L., Tzur, R., Hodkowski, N., Jorgensen, C., Wei, B., Wang, X., & Davis, A. (2018). A written, large-scale assessment measuring gradations in students' multiplicative reasoning. In E. Bergvist, M. Osterholm, C. Granberg, & L. Sumpter (Eds.), *Proceedings of the 42nd conference of the International Group for the Psychology of Mathematics Education* (Vol. 3, pp. 163–170). Umea, Sweden: PME.

Kane, M. T. (1992). An argument-based approach to validity. *Psychological Bulletin, 112*(3), 527.

Kane, M. T. (2012). Validating score interpretations and uses: Messick lecture, language testing research colloquium, Cambridge, April 2010. *Language Testing, 29*(1), 3–17.

Kloosterman, P., Mohr, D., & Walcott, C. (2016). *What mathematics do students know and how is that knowledge changing? Evidence from the National Assessment of Educational Progress.* Charlotte, NC: Information Age Publishing.

Kosko, K. W. (2019). A multiplicative reasoning assessment for fourth and fifth grade students. *Studies in Educational Evaluation, 60*, 32–42.

Kosko, K. W. (2018a). Reconsidering the role of disembedding in multiplicative concepts: Extending theory from the process of developing a quantitative measure. *Investigations in Mathematics Learning, 10*(1), 54–65.

Kosko, K. W. (2018b). *The multiplicative meaning conveyed by visual representations.* Manuscript under review.

Kosko, K. W., & Singh, R. (2018a). Elementary children's multiplicative reasoning: Initial validation of a written assessment. *The Mathematics Educator, 27*(1), 3–22.

Kosko, K. W., & Singh, R. (2018b). What form of mathematics are assessments assessing? The case of multiplication and division in fourth grade NAEP items. *Journal of Mathematics Education at Teachers College, 9*(1), 1–8.

Linacre J. M. (2010) When to stop removing items and persons in Rasch misfit analysis? *Rasch Measurement Transactions, 23*(4), 1241.

Mulligan, J. T., & Mitchelmore, M. C. (1997). Young children's intuitive models of multiplication and division. *Journal for Research in Mathematics Education*, 28(3), 309–330.

Nunes, T., Bryant, P., Barros, R., & Sylva, K. (2012). The relative importance of two different mathematical abilities to mathematical achievement. *British Journal of Educational Psychology*, 82(1), 136–156.

Nunes, T., Bryant, P., Evans, D., & Barros, R. (2015). Assessing quantitative reasoning in young children. *Mathematical Thinking and Learning*, 17, 178–196.

Organisation for Economic Cooperation and Development [OECD]. (2016). *Country note: Key findings from PISA 2015 for the United States* (Research Report). Retrieved from www.oecd.org/pisa/pisa-2015-United-States.pdf

Siegel, S., & Castellan, N. J. (1988). *Nonparametric statistics for the behavioral sciences* (2nd ed.). New York: McGraw-Hill, Inc.

Siemon, D., & Breed, M. (2006). *Assessing multiplicative thinking using rich tasks*. Paper presented at the annual Conference of the Australian Association for Research in Education.

Siemon, D., Izard, J., Breed, M., & Virgona, J. (2006). The derivation of a learning assessment framework for multiplicative thinking. In J. Novotna, H. Moraova, M. Kratka, & N. Stehlikova (Eds.), *Proceedings of the 30th conference of the International Group for the Psychology of Mathematics Education* (Vol. 5, pp. 113–120). Prague: PME.

Simes, R. J. (1986). An improved Bonferroni procedure for multiple tests of significance. *Biometrika*, 73(3), 751–754.

Smith, S. Z., & Smith, M. E. (2006). Assessing elementary understanding of multiplication concepts. *School Science and Mathematics*, 106(3), 140–149.

Steffe, L. P. (1991). Operations that generate quantity. *Learning and Individual Differences*, 3(1), 61–82.

Steffe, L. P. (1994). Children's multiplying schemes. In G. Harel & J. Confrey (Eds.), *The development of multiplicative reasoning in the learning of mathematics* (pp. 3–39). Albany, NY: State University of New York.

Tzur, R., Johnson, H. L., Norton, A., Davis, A., Wang, X., Ferrara, M., Jorgensen, C., & Wei, B. (2017). Conceptions of number as a composite unit predicts students' multiplicative reasoning: Quantitative corroboration of Steffe's model. In B. Kaur, W. K. Ho, T. L. Toh, & B. H. Choy (Eds.), *Proceedings of the 41st conference of the International Group for the Psychology of Mathematics Education* (Vol. 4, pp. 289–296). Singapore: PME.

Ulrich, C. (2015). Stages in constructing and coordinating units additively and multiplicatively (Part 1). *For the Learning of Mathematics*, 35(3), 2–7.

Walkington, C., Clinton, V., & Shivraj, P. (2018). How readability factors are differentially associated with performance for students of different backgrounds when solving mathematics word problems. *American Educational Research Journal*, 55(2), 362–414.

Wilson, M. (2010). *Constructing measures: An item response modeling approach*. New York, NY: Routledge.

3 Substantiating Claims About Students' Algebraic Reasoning

Initial Evidence Based on Response Processes and Internal Structure

Lindsey Perry, Leanne R. Ketterlin-Geller, and Cassandra Hatfield

Introduction

Algebra, often considered a "gatekeeper" course (Wu, 2001), is critical for future success in later mathematics courses and is predictive of post-secondary completion (National Mathematics Advisory Panel, 2008). Foundational to success in algebra is numeric relational reasoning, the ability to analyze relationships between expressions or numbers and use those relationships to solve a problem mentally with minimal calculations (Carpenter, Franke, & Levi, 2003; Farrington-Flint, Canobi, Wood, & Faulkner, 2007). However, many students struggle with numeric relational reasoning and exhibit numerous misconceptions when solving problems (e.g., Behr, Erlwanger, & Nichols, 1980; Falkner, Levi, & Carpenter, 1999; McNeil & Alibali, 2005a), including misconceptions that suggest students read equations from left to right, as opposed to reasoning about the equality of the expressions. This misconception may stem from an overrepresentation of standard equations (i.e., $a + b = c$) in the classroom (Carpenter et al., 2003; Molina & Ambrose, 2006). Overexposure to standard equations and a lack of emphasis on the equal sign as a relational symbol (Powell, 2012) may indicate to students that operations must be performed from left to right. This leads many students to conceptualize the equal sign as a "do something" symbol (Behr et al., 1980, p. 15) instead of a relational symbol. For example, in the equation $6 + 4 = \Box + 5$, many students say the answer is 10, since they believe that the equal sign signals them to "do something" or to complete the operation on the left side of the equation. These misconceptions prevent students from reasoning relationally when solving equations or inequalities and hinder their ability to see relationships between expressions or numbers.

While understanding students' abilities to reason relationally is important, it is difficult to capture students' reasoning processes on traditional paper-and-pencil tests. For example, the item $6 + 4 = \Box + 5$ may be designed to assess students' numeric relational reasoning (e.g., since 5 is one more than 4, the unknown must be one less than 6). However, without understanding students' response processes, claims cannot be made about their reasoning abilities based only on their written response. Students could solve the equation and not necessarily reason about the relationships between the expressions. Because of this, if interpretations are to be made about students' reasoning abilities from their scores, then additional evidence must be collected to substantiate those claims.

In this chapter, we discuss the Algebra Readiness Progress Monitoring (ARPM) system. The ARPM is a system of three subtests with multiple parallel forms that can be used to monitor students' progress (Ketterlin-Geller, Gifford, & Perry, 2015). On these subtests, students are asked to examine two numbers or expressions to determine whether they are equivalent or if one is greater than the other. These items are designed to elicit students' numeric relational reasoning abilities. As such, validity evidence to substantiate this claim is needed. Two studies were conducted to collect preliminary evidence regarding two primary claims about the items and students' reasoning skills: (1) items designed using a systematically developed template elicit the same reasoning strategy, and (2) items may be solved using numeric relational reasoning.

Additionally, previous research completed on reasoning with these topics (e.g., Falkner et al., 1999; McNeil & Alibali, 2005b; Stephens et al., 2013) has primarily focused on how students view and find a missing value in non-standard equations, such as $a + b = c + d$. Different from previous research, this chapter presents evidence about how students interact with equations and inequalities when they are asked to compare two expressions using a comparison symbol (i.e., =, <, or >). The evidence collected to support the claims outlined above also helps clarify whether students potentially still view equations or inequalities from left to right even in the absence of a missing value. This provides a basis for validity evidence based on response processes.

ARPM Instrument

Assessment Purpose

The ARPM system is designed to be used within a Multi-tiered System of Support (MTSS). The purpose of the ARPM system at grades 6–8 is to provide teachers with ongoing data that can be used to monitor the development of students' algebra readiness knowledge and skills throughout the academic year. The ARPM system can be administered to students who have already been identified on a universal screener as struggling and are receiving supplemental interventions (Hatfield et al., 2016). Three subtests were developed for the ARPM system: Quantity Discrimination, Number Properties, and Proportional Reasoning. Administering these three subtests throughout the school year can provide data to help teachers and administrators determine if students are making sufficient progress toward reaching their goals and are responding to the supplemental interventions as anticipated. Additionally, students can chart their own personal growth over time.

Assessment Framework

The assessed construct for the ARPM system was defined in a two-step process. First, evidence was examined from a systematic content review indicating important algebra readiness concepts and skills for the middle grades (Ketterlin-Geller & Chard, 2011; Ketterlin-Geller et al., 2015). Second, mathematics educators and researchers further analyzed these findings in conjunction with learning progressions and state and national standards. This analysis led to a consensus on the definition of each for the three subtests.

- Quantity Discrimination (QD): This subtest is used to identify students' ability to reason flexibly within number systems (e.g., whole numbers, integers, and rational numbers). The ability to reason flexibly within number systems requires conceptual understanding.
  (National Mathematics Advisory Panel, 2008)

- Number Properties (NP): This subtest is used to identify students' applications of properties of operations (i.e., distributive property, commutative, associative, identity, and inverse properties of addition and multiplication, and equality). Students' ability to apply properties of operations supports efficient problem solving, development of algorithms, and sets the stage for symbolic manipulation needed in algebra.
  (Geary et al., 2008)

- Proportional Reasoning (PR): This subtest is used to identify students' ability to generalize the use of number systems and properties to solve and compare proportions. This application should be done without reliance on procedural calculations.

Assessment Structure and Administration

The assessment is structured to support ongoing progress monitoring. At each grade level, for each subtest, there are 20 forms with 30 items on each form. This allows for frequent administration. As an example, a sixth grader who has been identified as needing supplemental interventions based on the results of a universal screener can be given the three Grade 6 subtests up to 20 times within the academic year to monitor his or her progress. At minimum, the student should be given all three subtests once each month (Gersten et al., 2009), and if needed, the three ARPM subtests can be administered in one sitting.

In total, the administration of the three subtests takes 9 minutes. Students are provided 3 minutes to answer as many of the 30 items within a subtest as possible. Each subtest is timed to encourage students to answer as many items as possible as quickly as they can. Because the assessment is meant to test reasoning, if students are taking a long time to respond to each item, it is likely that they are not using reasoning but instead are calculating.

Each item is formatted in the same way: a numerical expression or statement on the left and the right, with a box in the middle. Students use selected response to identify which symbol (<, =, or >) makes the statement about the two numerical expressions or values true. Each item is designed in such a way that students can use numeric relational reasoning between the numbers or expressions to make the comparison. For example, to solve $3\frac{5}{9}+1\frac{3}{4}$ $\square$ $1\frac{3}{4}+8\frac{5}{9}$, students could reason that because of the commutative property of addition, the order of the addends does not affect the sum. Thus, the comparison $a + b$ $\square$ $b + c$ can be determined by comparing a and c, or in this case, comparing $3\frac{5}{9}$ and $8\frac{5}{9}$.

Monitoring progress necessitates parallel forms; making inferences about student growth based on changes in performance requires that the forms be parallel in nature. Parallel forms should have very similar test information functions and test characteristics (e.g., test length) (Lord, 1980). To achieve this objective, we focused on the parallel nature of the items. An item template was designed to help create items that were as similar as possible in reasoning, difficulty, and content.

The item template included an exemplar item and a generalized item model in algebraic terms to describe the exemplar item. The exemplar item was used for the first form of each subtest and the item model was used to develop the items for forms 2–20. Within the item model, constraints were used to limit the values on each variable. As an example, 2^3 $\square$ 2×3 was designed as an exemplar item for Quantity Discrimination. The generalized item model was d^a $\square$ $d \times a$, where $2 \le d \le 9$ and $3 \le a \le 9$. The item template also included three sample items created using the generalized item model. As a result of using this method, parallel items were written that were similar in content, hypothesized reasoning strategy, and difficulty level. By focusing on using the item template to create items that are parallel in nature, the resulting ARPM forms should also be parallel.

Validity

There are two primary claims that support the intended interpretations and uses of the ARPM system: (1) items designed using a systematically developed template elicit the same reasoning strategy, and (2) items may be solved using numeric relational reasoning. These claims are part of a larger interpretation-use argument (Kane, 2013), discussed below. These claims rest on assumptions about students' response processes and on the technical adequacy of the items. This section provides an overview of validity and the two sources of validity evidence that align with the claims outlined above.

Validity refers to the meaningfulness and trustworthiness of the interpretations and uses of test scores and is an iterative process (American Educational Research Association, American Psychological Association, & National Council on Measurement in Education, 2014). To infer validity, test developers need to provide sufficiently convincing evidence that justifies the claims about test score meaning. Alternative hypotheses should be examined and counterarguments refuted. As such, validation requires a minimum of three critical components: (1) a series of claims about the meaning of the test scores, (2) a collection of evidence examining the plausibility of the claims, and (3) an evaluation of the evidence against the claims to examine their credibility and support the intended interpretations and uses of the test scores (Kane, 2006, 2013).

In the context of validation, claims about score meaning may be organized into an argument-based framework (Kane, 2006, 2013). Within this framing of validity, claims are assertions about the meaning of the observed performance and are typically proposed by the test developers. These claims should represent a coherent and comprehensive case that connects—in sufficient detail—the observed performance with the intended interpretations and uses of the test scores. These claims provide the basis from which relevant sources of evidence are identified and subsequently collected. The sources of evidence should align with the claims both in terms of intention and causality; in other words, claims should direct the types of data (e.g., theoretical or empirical) and the certainty of the conclusions that are needed to support the intended interpretations and uses of the test scores. The *Standards for Educational and Psychological Testing* (American Educational Research Association et al., 2014), subsequently referred to as the *Test Standards*, provide test developers with guidance on plausible sources of evidence that may support specific claims. It is the test developers' responsibility to identify sources of evidence that are relevant for the claims being proposed.

Five sources of evidence are suggested by the *Test Standards* (American Educational Research Association et al., 2014): test content, response processes, internal structure, relation to other variables, and consequential aspects of testing. Although a thorough treatment of each source of evidence is beyond the scope of this chapter, evidence based on test content is defined as the correspondence between the tested content and the intended construct. Evidence based on response processes is defined as the alignment between the thoughts, behaviors, or actions that are elicited when the examinee responds to an item and the intended thoughts, behaviors, or actions. Evidence based on internal structure is defined as the association between the item and test components that underlie an examinee's score and the intended construct. Evidence based on relation to other variables seeks to examine the correspondence between an examinee's performance across multiple measures of similar or dissimilar constructs. Finally, evidence for validity and consequences of testing seeks to examine the decisions and associated outcomes that precipitate from test score use.

An interpretation-use argument (IUA) (Kane, 2013) can be created to organize and outline the claims and sources of evidence that link a test score to its intended interpretation. A portion of the IUA for the claims discussed in this chapter can be seen in Figure 3.1. The two claims focused on for this chapter are that (1) items designed using a systematically developed template elicit the same reasoning strategy, and (2) items may be solved using numeric relational reasoning. The IUA for the ARPM system identifies these claims and the evidence collected to provide initial evidence to substantiate these claims. Additional claims would be needed to fully link the test score and the intended interpretation. For example, since the ARPM system is intended to generate data for monitoring progress, an additional claim could be that the forms designed to be parallel in nature have similar content and difficulty levels. In this chapter, we focus on the

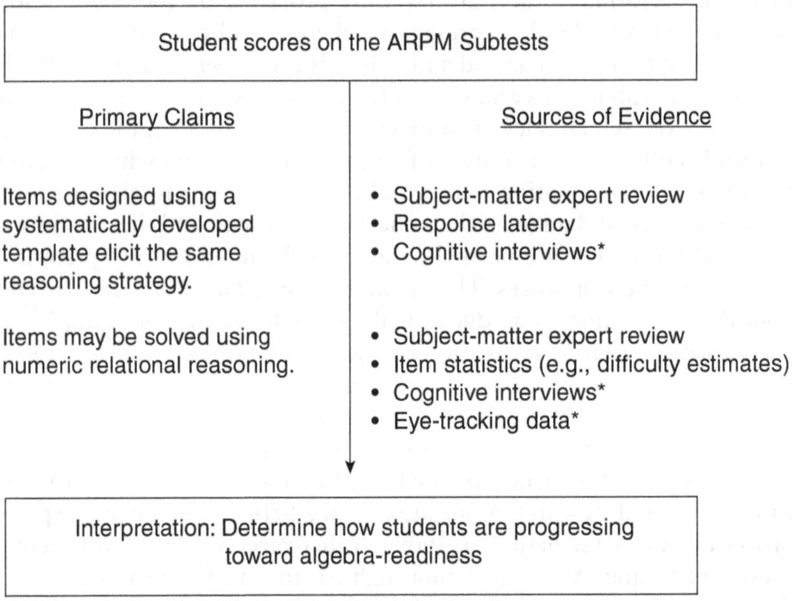

Figure 3.1 Part of the interpretation-use argument for the Algebra Readiness Progress Monitoring (ARPM) system

*Possible sources of evidence not contained in this chapter.

two claims identified on the IUA and the associated sources of evidence related to the response processes and internal structure of the ARPM.

Evidence Based on Response Processes

The *Test Standards* (American Educational Research Association et al., 2014) define evidence based on responses processes as theoretical or empirical data that links the construct with the observed responding behaviors of the examinee. Hubley and Zumbo (2017) operationalize response processes as "the mechanism that underlies what people do, think, or feel when interacting with, and responding to, the item or task and are responsible for generating observe test score variation" (p. 2). Based on this definition, response processes extend beyond the cognitive aspects of responding to include affective, motivational, and emotional mechanisms that influence the ways in which an examinee attends to and responds to an item (Leighton, Tang, & Guo, 2017). Evidence associated with responses processes are most common for claims that explicitly associate the observed performance with specific cognitive processes (e.g., reasoning, problem solving) (American Educational Research Association et al., 2014); however, the data gathered about response processes can be used to better understand the examinee's responding behavior for any assessment.

Evidence examining test takers' response processes can be used for several validity-related claims. First, response processes data can be used to examine claims that examinees are interpreting item content and/or characteristics or features as intended by the item writer. These claims are fundamental to all testing programs, including surveys (Padilla & Leighton, 2017). In some cases, evidence may indicate that there are construct-irrelevant aspects of items that influence examinees' responding behavior (Leighton et al., 2017), which can be addressed by revising or omitting the items. Second, response processing data can be used to examine the correspondence between the cognitive processes underlying the construct definition and those that are actually being elicited as examinees attend to and respond to the items (Embretson & Gorin, 2001).

Until recently, limited attention has been placed on examinees' response processing, and few studies have focused on the depth and breadth of response processing data (Ercikan & Pellegrino, 2017; Hubley & Zumbo, 2017). As a result, the accepted sources of evidence to investigate examinees' response processes are currently limited in range, but advances are rapidly being made. Possibly the most widely collected evidence is verbal reports or descriptions obtained directly from the examinee. These data—often captured using cognitive interview or think-aloud interview techniques—provide direct insights into the examinee's thinking by probing them to reflect on or actively narrate their responding behavior. Some concerns have been raised about the overreliance on verbal reports because they may cause unintended stress or anxiety on the examinee

(Leighton et al., 2017) and may introduce artificial processes outside of those required to respond to the item (e.g., thinking aloud while solving a problem). Moreover, as with all qualitative data, the ability to generalize findings to the broader population is not possible. As such, researchers have called for advances in methods for examining response processes (Ercikan & Pellegrino, 2017; Hubley & Zumbo, 2017).

Other data sources are emerging that may contribute to validity evidence underlying examinees' response processing. Neural processing data such as response latency or eye movements provide evidence about the extent of the cognitive processes being engaged during responding, but can also be used to detect aberrant behavior when an examinee's pattern of responding varies from the inter- or intra-examinee expected behavior (Li, Banerjee, & Zumbo, 2017; Oranje, Gorin, Jia, & Kerr, 2017). Response latency (e.g., the time lapse between presentation of the items and response) may be useful for examining both cognitive (e.g., cognitive load) and non-cognitive (e.g., level of attention) factors involved in responding to items (Li et al., 2017). Physiological data (e.g., heart rate, sweat production, hormone levels) may be used to infer affective or emotional states such as stress or anxiety, especially when changes are detected (Leighton et al., 2017). These evidence sources can be applied directly to the examinee or to a rater who is generating a score (Hubley & Zumbo, 2017).

For ARPM, the two primary claims focused on in this chapter center on students' hypothesized and actual response processes. The first claim is that items designed using the template elicit the same reasoning strategy. Because the ARPM system has multiple forms that can be given to monitor progress, it is important that the items are parallel in nature, not only in content and difficulty level but also in hypothesized reasoning strategy. Because the items claim to assess students' reasoning, evidence must be collected to substantiate that claim. The second claim is that items may be solved using numeric relational reasoning. Since numeric relational reasoning is the construct being assessed, collecting evidence about how students process the items is critical for construct-related evidence of validity.

Evidence Based on Internal Structure

Internal structure of an instrument refers to the technical characteristics that underlie the generation of an examinee's score, such as the item and test scoring procedures, dimensionality of behaviors, and sources and impact of variability among examinees (Rios & Wells, 2014). Claims that link the technical characteristics with the assessed construct and intended interpretations and uses are commonly made by test developers and underlie the reliability of the scores. Within an argument-based framework of validation, these claims are often associated with scoring inferences, which purport correspondence between the examinees' observed performance and their scores and seek to determine if the scores are appropriate, accurate, and consistent (Kane, 2006, 2013).

Because of the omnipresent nature of claims asserting the internal structure of tests, there are numerous plausible sources of validity evidence. Standards 1.13–1.15 from the *Test Standards* identify three key claims about the internal structure of tests and propose possible sources of evidence. These claims include assertions about (1) the relationship among items or parts of the test including the dimensionality of the behaviors, which can be supported with evidence based on the factor structure; (2) the meaning of subscores, score profiles, and/or composite scores, which can be supported with evidence about the distinctiveness and reliability of the scores; and (3) the confidence the test user can reasonably have for interpreting and using the data for specific decisions, which can be supported with documentation of the error for specific items, subsets of items, or subscores (American Educational Research Association et al., 2014).

Evidence of internal structure is also closely related to the two primary reasoning claims about the ARPM items described in this chapter. Interpretations made using scores from items without sufficient technical adequacy data may not be valid and must be interpreted with caution. In order to make claims about students' reasoning using the ARPM items, the items must be technically adequate.

ARPM Response Process Evidence

In this section, two studies are detailed that provide different sources of evidence about response processes and internal structure to help substantiate the two primary claims: (1) items designed using a systematically developed template elicit the same reasoning strategy, and (2) items may be solved using numeric relational reasoning. Study 1 involved a comprehensive review of the item templates by subject-matter experts to ensure that the items created using the item template to be parallel in nature could be solved by using the same numerical reasoning strategy, were aligned with the content, and had hypothetically similar difficulty levels. The data from this study provide initial evidence for both claims. Study 2 involves response process evidence from a field test of the ARPM items for the Number Properties and Proportional Reasoning subtests. This study focused on determining if students responded differently to items based on the placement of expressions. Item parameters and response latency were examined. The evidence from Study 2 provides preliminary support for the second claim.

Study 1: Subject-Matter Expert Review

After an iterative cycle of item writing and internal review, a subset of 25% of the item templates were reviewed by two subject-matter experts to provide evidence of content alignment, hypothesized response processes, estimated item difficulty, and accuracy.

Participants

Two subject-matter experts participated in the item review. Both experts have doctoral and master's degrees in mathematics education and curriculum and instruction. One subject-matter expert is a professor emerita in a department of mathematics at a large state university. This expert has been heavily involved in mathematics education at the state and national levels and has been instrumental in helping publish curriculum standards and multiple practitioner resources for grades K-8. The second expert is a research professor in mathematics education and has published a multitude of mathematics education books and articles. This expert has been the co-principal investigator for two large algebraic reasoning projects.

Methods

The two subject-matter experts were asked to review each item template, which included the exemplar item, the generalized item model, and three additional items created using the item model described above. The reviewers responded to four statements about each item: (1) the targeted category and subcategory is assessed; 2) the four items assess the skills at a similar level of difficulty; 3) a similar numerical reasoning strategy can be used on all four items; and 4) the intended correct answer is true for all four items. The third statement provides evidence about students' hypothesized response processes. As previously mentioned, one of the primary claims for the ARPM system is that the items designed using the template elicit the same reasoning strategy. Evidence from this question aligns with this claim.

Results

Percentages were calculated to summarize the reviewers' ratings. Reviewer 1 and Reviewer 2 agreed that a similar numeric relational reasoning strategy could be used on all four items for most of the item templates: 92% and 92% for Quantity Discrimination, 100% and 88.9% for Proportional Reasoning, and 100% and 87% for Number Properties. Reviewer 1 and Reviewer 2 also agreed that the targeted category and subcategory were assessed: 100% and 100% on Quantity Discrimination, 87% and 90% on Number Properties, and 100% and 96.3% on Proportional Reasoning. Additionally, the reviewers noted that the four items on each template assessed the skills at a similar difficulty level: 92% and 85% for Quantity Discrimination, 100% and 84% for Number Properties, and 67% and 74.1% for Proportional Reasoning. They also noted that the

Table 3.1 Example Items and the Accompanying "Flipped" Item from the Number Properties and Proportional Reasoning Subtests

	Grade	Item Location	Item
Number Properties Subtest	6	10	$(200 + 40)3 \; \square \; 600 + 120$
		37	$600 + 120 \; \square \; (200 + 40)3$
	7	26	$2 + 4 \; \square \; 7\left(\dfrac{2}{7} + \dfrac{4}{7}\right)$
		39	$7\left(\dfrac{2}{7} + \dfrac{4}{7}\right) \; \square \; 2 + 4$
	8	8	$-3\dfrac{1}{4} \times \left(5\dfrac{5}{6} \times 4\right) \; \square \; \left(-3\dfrac{3}{4} \times 5\dfrac{5}{6}\right) \times 4$
		36	$\left(-3\dfrac{3}{4} \times 5\dfrac{5}{6}\right) \times 4 \; \square \; -3\dfrac{1}{4} \times \left(5\dfrac{5}{6} \times 4\right)$
Proportional Reasoning Subtest	6	27	$\dfrac{2}{5}$ of 18 $\square$ $\dfrac{2}{5}$ of 25
		39	$\dfrac{2}{5}$ of 25 $\square$ $\dfrac{2}{5}$ of 18
	7	3	46% of 49 $\square$ 46% of 39
		17	46% of 39 $\square$ 46% of 49
	8	5	$\dfrac{1}{7}$ of 28 $\square$ $\dfrac{11}{6}$ of 28
		36	$\dfrac{11}{6}$ of 28 $\square$ $\dfrac{1}{7}$ of 28

correct answer was true for all four items: 96% and 100% for Quantity Discrimination, 97% and 100% for Number Properties, and 100% and 100% for Proportional Reasoning. If the reviewers disagreed with any of the statements, changes were made to improve the items based on their feedback.

Study 2: Flipped Items

To investigate the second claim about whether the items elicit numeric relational reasoning, a total of 13 of the Number Properties items and 14 of the Proportional Reasoning items from grades 6–8 were selected to be presented to students as "flipped" items. Students responded to the original item but also responded to an item that shifted or "flipped" the placement of the expressions. In other words, the expression on the left side of the original item was placed on the right side of the comparison box in the new item. Table 3.1 includes examples of the original and "flipped" items. The items selected to be "flipped" were chosen based on their content (see Table 3.2 for a breakdown of the "flipped" items by grade for each subtest) and based on an anticipated response process. For example, some items were selected because they could be solved very differently if a student read the expressions from left to right versus, examining how the two expressions were related. Additionally, only a few items per grade level were chosen as to not unduly lengthen the assessment.

Students were administered the "flipped" items as well as the original 35 items per subtest in an online testing environment under the direction of a trained assessor. Students were not allowed to go backward to revise answers. Each of the 13 "flipped" items was placed an average distance of 18 items ($SD = 9$) away from its respective original item. It was important to adequately space the original and "flipped" items to prevent students from remembering their first response to the item. Spacing out the items was intended to ensure that students did not cue off the original item prior to responding to the "flipped" item. The online testing environment also tracked the number of seconds spent on each item.

Participants

Students in grades 6–8 from a high-achieving middle school in Texas participated in this study. All students from six teachers' classrooms (i.e., two teachers per grade level) were recruited to participate. After gaining parental consent and student assent, 39–59 students per grade level took each subtest (see Table 3.2 for the exact number of participants). Most of the students were white (between 89% and 95% per grade level per subtest) and female (between 55% and 70% per grade level per subtest). This study was approved by the authors' institutional review board.

Analyses

Student responses were scored dichotomously, and responses to each subtest were analyzed using the Rasch model. The subtests were assumed to be unidimensional based on the evidence from Study 1; however, unidimensionality was not confirmed analytically. The Rasch model estimates each item's difficulty level (b-parameter) and the probability that a person will respond to an item correctly. The item parameters can be compared to see if differences exist between items (e.g., the original and "flipped" items). The typical range for b-parameters is $-3 \leq b \leq 3$ (Baker, 2001), where the scale goes from less difficult (negative b-parameter) to more difficult (positive b-parameter). Winsteps (Linacre, 2012) was used to conduct these analyses. Item fit statistics, such as point measure (i.e., the correlation between the scores on observations and the scores on the entire measure) and mean-squares infit and outfit estimates, were also calculated to investigate the technical adequacy of the items. Infit is

Table 3.2 Number of Original Items per Grade Level for the Number Properties (NP) and Proportional Reasoning (PR) Subtests

NP Subtest	Grade	Number of students	Number of Original Items				
			Distributive Property	Commutative Property	Associative Property	Identity: Multiplicative	Identity: Addition
	6	42	2	1	1	1	
	7	52	2		1		1
	8	39	1	1	1		1

PR Subtest	Grade	Number of students	Number of Original Items			
			Fraction comparison	Unit fraction of a whole number	Fraction of a whole number	Percent of a whole number
	6	43	3	1	1	
	7	59	1	1	1	2
	8	39		1	1	2

sensitive to inliers and outfit is sensitive to outliers. Each examines patterns of responses to items with difficulty levels near or far from examinees' ability levels, and values of 0.5–1.5 are considered "productive for measurement" (Linacre, 2002). For Rasch, it is preferred to have samples of at least 50 (Linacre, 1994). In this study, not all grade levels and subtests had over 50 participants. This may impact the stability of some of the estimates. To investigate differences between the item difficulty estimates of the original and "flipped" items, the difference between the item difficulty of each original item and its respective "flipped" item was calculated.

Additionally, differences in response times between the original and "flipped" items were calculated to determine if there were differences in how quickly students responded to the original and "flipped" items. Large differences in response times might indicate that students are using different strategies to solve the original and "flipped" items.

Results

Item Difficulty Estimates

The item difficulty estimates and item fit statistics for the original and "flipped" items are shown in Table 3.3. Overall, both the original and "flipped" items had satisfactory item fit statistics, using the criteria stated above. Most items had mean-square infit values in the suggested range of 0.5–1.5 (Linacre, 2012). All but two Number Property items and four Proportional Reasoning items had mean-square outfit values in this range. All Number Property items and all but four Proportional Reasoning items had point measure estimates above 0.25. As such, the item fit statistics for the "flipped" items were not markedly different from item fit statistics for the original items. The item difficulty estimates for the original and "flipped" items as a whole were also not markedly different (see Table 3.3).

While the overall item difficulty estimates for the original and "flipped" items are relatively similar, large differences exist between the item difficulty estimates for the original and "flipped" items *at the item level* by property or type. Table 3.4 shows the mean difference between item difficulty estimates for the original and "flipped" items by grade level and by property or item type.

Response Latency

The pilot test was untimed; however, lapsed time was collected at the item level. Students took an average of 37 seconds ($SD = 21$), 20 seconds ($SD = 12$), and 25 seconds ($SD = 12$) for each grade 6, grade 7, and grade 8 Number Property item, respectively. Students took an average of 19 seconds ($SD = 7$), 15 seconds ($SD = 4$), and 18 seconds ($SD = 5$) for each grade 6, grade 7, and grade 8 Proportional Reasoning item, respectively.

Table 3.3 Item Statistics for Original and "Flipped" Items

	Number Properties Subtest						Proportional Reasoning Subtest					
	Grade 6		Grade 7		Grade 8		Grade 6		Grade 7		Grade 8	
	Original	Flipped	Original	Flipped	Original	Flipped	Original	Flipped	Original	Flipped	Original	Flipped
Average item difficulty (SD)	0.13 (1.14)	0.02 (0.66)	0.11 (2.21)	0.17 (2.50)	0.33 (1.45)	0.48 (1.00)	-0.62 (1.71)	-0.30 (1.46)	-0.04 (1.28)	0.02 (0.91)	0.18 (1.22)	-0.38 (1.28)
Average mean-square infit (SD)	1.08 (0.09)	1.04 (0.15)	0.94 (0.12)	0.94 (0.16)	1.05 (0.20)	0.89 (0.07)	0.96 (0.17)	0.99 (0.12)	0.99 (0.12)	0.90 (0.20)	1.00 (0.14)	0.88 (0.06)
Average mean-square outfit (SD)	1.05 (0.14)	0.99 (0.29)	0.89 (0.18)	1.14 (0.96)	1.10 (0.41)	0.81 (0.12)	0.76 (0.30)	1.06 (0.45)	1.98 (1.98)	0.98 (0.45)	1.45 (1.07)	1.05 (0.62)
Average point measure (SD)	0.28 (0.08)	0.28 (0.05)	0.25 (0.04)	0.24 (0.03)	0.34 (0.03)	0.37 (0.01)	0.29 (0.12)	0.30 (0.07)	0.36 (0.09)	0.38 (0.06)	0.36 (0.07)	0.32 (0.09)

Table 3.4 Mean Difference Between Item Difficulty Parameters for Original Items and "Flipped" Items (SD) on the Number Properties Subtest

			Mean difference between item difficulty (SD)
Number Property Subtest	Grade	6	0.76 (0.40)
		7	0.52 (0.30)
		8	0.43 (0.44)
	Property	Distributive	0.68 (0.27)
		Commutative	0.60 (0.69)
		Associative	0.52 (0.36)
		Inverse: Multiplicative	1.12
		Identity: Addition	0.15 (0.05)
Proportional Reasoning Subtest	Grade	6	0.80 (0.42)
		7	0.44 (0.24)
		8	0.55 (0.18)
	Item Type	Fraction comparison	0.72 (0.29)
		Unit fraction of a whole number	0.43 (0.23)
		Fraction of a whole number	0.69 (0.38)
		Percent of a whole number	0.37 (0.44)

Differences were detected between the response latency for the original items and the "flipped" items. For the Number Property items, there were statistically significant differences ($p < 0.05$) found between the response times for every pair of items (i.e., the original and the "flipped" item) at every grade level. In other words, students took on average a different amount of time on the original item compared to the "flipped" item. Differences were also seen with the Proportional Reasoning items. However, not all pairs of items had statistically significant differences; 2 out of 4, 4 out of 5, and 3 out of 4 items at grade 6, 7, and 8 had statistically significant differences ($p < 0.05$) between the response times for the original and "flipped" items. It is important to note that for every pair of items except for two grade 6 Proportional Reasoning items, students took longer to respond to the original item, which appeared first. This may indicate that as students completed the assessment they realized strategies they could use to solve the problems more quickly. However, even though the students answered almost all of the "flipped" items more quickly, which were presented later in the test than the original items, the item difficulty levels did not indicate that students were more accurate with the "flipped" items. The percentages of item pairs that had a faster response rate on the "flipped" item but a higher difficulty level on the "flipped" item were 40%, 50%, and 50% for grades 6, 7, and 8 for Number Properties and 20%, 40%, and 0% for grades 6, 7, and 8 for Proportional Reasoning.

Discussion

The purpose of this chapter was to describe the ARPM system and present initial validity evidence gathered to examine the trustworthiness and meaningfulness of two claims on the IUA (see Figure 3.1): (1) items designed using a systematically developed template elicit the same reasoning strategy, and (2) items may be solved using numeric relational reasoning. Both of these claims impact the validity of the interpretations made using this assessment system. The ARPM system was designed to offer teachers and schools a tool that can provide ongoing data about students' progress in learning algebra readiness knowledge and skills. Since the interpretations being made using the ARPM system focus heavily on students' algebra readiness and numeric relational reasoning skills, it is vital that there is evidence supporting these two claims. This chapter provides initial evidence that can be used to begin to substantiate these claims. This section evaluates the claims in light of the evidence collected and presents information on additional methods that can be used to collect evidence based on response processes: interviews and eye-tracking studies.

Evidence Supporting Claims About Students' Reasoning

Claim 1: Items Designed Using a Systematically Developed Template Elicit the Same Reasoning Strategy

The first claim on the IUA focuses on whether the items developed for the ARPM system elicit the same reasoning strategy. As previously noted, the ARPM system has three subtests: Quantity Discrimination, Number Properties, and Proportional Reasoning. Each subtest has 20 forms that are designed to be parallel in nature. An exemplar item was created and a generalized item model was created to describe each exemplar item in algebraic terms. This item model was used to create the items for all forms for each subtest. This process was enacted to control for the content, reasoning strategy, and estimated level of difficulty across forms. The subject-matter experts who reviewed the items in Study 1 noted that a large majority of the items designed following the item templates were parallel in nature based on reasoning strategy, difficulty level, and content. The reviewers came to this conclusion after reviewing the exemplar item, the generalized item model, and three additional items written using the generalized item model as a guide.

The ratings from the subject-matter experts suggest that the items designed using the item templates to be parallel in nature do indeed elicit the same reasoning strategy and would likely result in similar response processes. If students examine the relationships between the expressions or quantities, they will likely use a similar strategy. For example, consider items of the form $a + b = b + a$, where a is a whole number greater than

10 and less than 20, and *b* is a whole number greater than 0 and less than 11. Students will likely either read the expressions from left to right and try to calculate the sums or use numeric relational reasoning to recognize that the expressions are equivalent because of the commutative property.

However, it is important to note limitations to this evidence. While the subject-matter experts rated the similarity of the items highly in their review, their ratings are based on how they perceive students will respond. Students may use different reasoning strategies for all of the items or different strategies for similar items across forms. The hypothesized response processes are similar across items designed to be parallel in nature; however, additional evidence is needed to empirically evaluate this claim.

Claim 2: Items May Be Solved Using Numeric Relational Reasoning

The second claim focuses on the idea that the ARPM items may be solved using numeric relational reasoning. The results from the Rasch analyses and response latency analyses suggest that students may not be consistently using numeric relational reasoning strategies to solve the items. In Study 2, we examined whether original items and "flipped" items had different item difficulty parameters and response times. If the original and "flipped" items had similar difficulty parameters and response times, it could be assumed that students were solving the items similarly and using the same strategy for both of the items. That could mean that students were consistently applying or not applying reasoning strategies. However, the item difficulty estimates for both subtests examined (i.e., Number Properties and Proportional Reasoning) indicate that there are differences between item difficulty parameters and response times. This means that students had different likelihoods of answering the original and "flipped" items correctly, which signifies that students may have been using different reasoning strategies to answer these items. For example, the items 2.8 + 6 □ 0.2(1.4 + 3) and 0.2(1.4 + 3) □ 2.8 + 6 had *b*-parameters of 0.98 and 0.08, respectively. This indicates that the item with the parentheses on the left side was less difficult; students may have started solving the expression on the left and realized that after applying the distributive property the values were the same as those in the expression on the right side. The item with the parentheses on the right side may have been more difficult because students read the item from left to right and immediately started calculating the expression on the left, without applying their numeric relational reasoning skills.

The magnitude of the differences between item difficulty estimates seems to depend on the grade level and the topic assessed (see Table 3.4). On the Number Properties subtest, as the grade level increased, the difference in item difficulty estimates decreased. On the Proportional Reasoning subtest, the difference in item difficulty estimates was greatest at

Grade 6. These differences may indicate that as students gain more exposure to equations and inequalities in middle school, they are more likely to focus on the relationships between expressions instead of the placement of the expressions. For example, students who read equations from left to right and do not focus on finding relationships between expressions may see the original and "flipped" items as two distinct items, which may potentially lead to different strategies, different responses, and different item difficulty estimates. However, students who see the relationship between the two expressions are likely to respond similarly to both items, which would result in similar item difficulty estimates.

Additionally, differences also appear to depend on the topic assessed. For the Number Properties subtest, certain properties, such as the distributive property, have larger differences in item difficulty estimates than other properties, such as the identity property of addition. These differences may be due to the types of calculations involved in the equations or inequalities and whether students are solving these items from left to right. For example, items assessing the identity property of addition (e.g., $0.45 + 0 \square 0.045$; $0.045 \square 0.45 + 0$) contain fewer numbers within each item and do not include parentheses, as are included in items assessing the distributive and associative properties (see Table 3.4). These types of items are more conducive to calculation, and solving from left to right may not hinder students' problem-solving processes. Therefore, it is logical that the differences in item difficulty estimates were smaller in the items assessing the identity property of addition compared to other properties.

More complex items, such as those assessing the distributive property, may not be as conducive to solving from left to right, which may have led to the marked differences in item difficulty estimates. For example, in the original item in Table 3.4, students may find the original item easier if they are working from left to right. They can solve $(200 + 40)3$, which results in the expression on the right. No addition is needed if they recognize this relationship. However, if students work from left to right on the "flipped" item, it may be more difficult to find the sum of $600 + 120$, keep that sum in working memory, and solve $(200 + 40)3$ completely. The item difficulty estimates indicate that the original item was indeed easier than the "flipped" item. This may suggest that students did not examine the relationship between expressions and were attempting to work the item from left to right. However, since these conclusions come from small samples of items, these are hypotheses that cannot yet be generalized. Additional data are needed to substantiate these conclusions.

These results suggest that differences in item difficulty estimates do exist between the original items and the "flipped" items. Put another way, the placement of expressions within an equation or inequality seems to impact the likelihood that a student will get the answer correct. These analyses, albeit limited because of a small sample size, suggest that students may be not be consistently applying numeric relational reasoning to solve these items. Linking back to the second claim on the IUA that these items can be solved using numeric relational reasoning, this evidence suggests that while these items can be solved using numeric relational reasoning, students are not consistently applying numeric relational reasoning strategies to these items. However, since this instrument is meant to help assess students' reasoning, it is appropriate that not all students immediately reason similarly on these items.

Additional Methods to Collect Evidence Based on Response Processes

While these results reveal insights into how students interact with equations and expressions, there are limitations to the claims that can be made about students' response processes based on these results. For example, the subject-matter experts noted whether the same reasoning strategy could be used on the items designed to be parallel in nature. While their ratings provide initial evidence for the first claim on the IUA and suggest that similar reasoning strategies could be used, this evidence is hypothetical in nature. Only evidence from direct interactions with students can indicate what reasoning strategies are being used on these items. Additionally, while the evidence presented in Study 2 indicates that students may not be consistently using numeric relational reasoning strategies and may be reading expressions from left to right, this hypothesis cannot be confirmed without evidence from other sources, such as interviews or eye-tracking studies. In this section, we discuss how interviews and eye-tracking could be used to collect additional evidence based on response processes.

Interviews

Cognitive interviews and think-aloud interviews can provide test developers with insight into students' thought processes as they solve or interact with an item. Cognitive interviews are interactions with test takers that examine their response processes to questions or items from a cognitive framework (Miller, Chepp, Wilson, & Padilla, 2014). Think-aloud interviews ask students to verbalize their thoughts as they respond to an item on an instrument (Ericsson & Simon, 1993). While both of these types of interviews ask students questions about their thinking, the questions within cognitive interviews are designed to bring out and examine students' cognition, whereas assessment items are typically used in think-aloud interviews and the data are used to refine the items.

Think-aloud interviews, with additional follow-up questions to probe for students' reasoning, could be used to collect evidence of students' response processes on the ARPM items. Think-aloud interviews typically consist of both concurrent and retrospective data collection elements. For the concurrent element, a student is asked to verbalize all of their thoughts and everything they are doing when solving an item. This verbalization is unfiltered and may not be linear in nature; it is meant to be a stream of consciousness. The interviewer does not ask any questions of the student while they are solving the item; however, the interviewer can prompt the student to "please think aloud" if the student remains silent for over a few seconds. This concurrent element is meant to provide insight into how the student solved the item, including what strategies he or she used and possibly what misconceptions he or she exhibited. The retrospective element of the think-aloud interview allows the researcher to ask predetermined questions about the item (e.g., What is the question asking you to do? What steps did you take to solve this problem? Was anything confusing or difficult about the item?) and also follow up on the actual responses from the concurrent think-aloud interview (e.g., Can you tell me more about how you ____?).

Using both concurrent and retrospective elements may be a beneficial way of collecting response process data for the ARPM subtests to strengthen the validity evidence for the two primary claims presented in this chapter. For example, consider the item $\frac{7}{9}+\frac{1}{3} \; \Box \; \frac{1}{10}+\frac{7}{9}$. Within the concurrent think-aloud interview, much can be learned about a students' reasoning process within the first few seconds of their verbalization. If they start out by saying "Ok, so seven-ninths plus one-third. I need to find a common denominator so I can add these fractions," it indicates that the student was reading the expressions from left to right and did not examine the relationships between the two expressions in the item. Instead, if the student said, "I have seven-ninths plus one-third on this side. I do see that there is also a seven-ninths on the other side. Since there is a seven-ninths on both sides, I can just compare the one-third and the one-tenth," it is evidence that they looked at both expressions, saw the similarities between the two expressions, and applied their knowledge of the lawfulness of the commutative property to reason through the item.

All of the data from the concurrent and retrospective elements can be transcribed and analyzed for themes. These themes may be related to different reasoning strategies or may be aligned with possible student misconceptions. While the ARPM items were designed to elicit numeric relational reasoning, conducting think-aloud interviews would provide additional evidence for this claim.

Eye-Tracking

Eye-tracking can also be a useful research method for collecting data on students' response processes. The eye-mind assumption (Just & Carpenter,

1980) suggests that eye movements indicate what a student is paying attention to or focusing on. This method has been commonly used in psychology to study cognition and information processing, and it allows researchers to directly observe cognitive processes more objectively than self-reports from participants. Eye-tracking experiments allow researchers to collect data on participants' eye fixations and saccades. Fixations are times and locations of stable eye movement, and saccades are the movements between two fixations. There are numerous measurements that can be collected and analyzed with eye-tracking including fixation duration, scanpath pattern, and saccade count (see Lai et al., 2013).

While research on reading has utilized eye-tracking studies for a few decades, as noted by Rayner (1998), eye-tracking studies in mathematics education have steadily increased in recent years. One study by Chesney, McNeil, Brockmole, and Kelley (2013) serves as an example of how eye-tracking methods could be used to investigate the ARPM items. To study whether an operational schema persists beyond elementary school, Chesney et al. (2013) investigated undergraduate students' eye movements as they solved equations of the form $a + b + c = d + \_\_\_$. These equations can be solved using numeric relational reasoning, but as noted previously, many children use an operational schema to solve these types of non-standard equations (e.g., add all of the numbers, add the numbers on the left side of the equal sign). With the eye-tracking data collected, they counted the number of times each students' gaze crossed over the equal sign from one expression to the other. Chesney et al. (2013) hypothesized that the number of back-and-forth eye movements would be predictive of using a relational reasoning strategy to solve the equation, and the results supported this hypothesis.

Eye-tracking methods could be used to collect additional evidence to support the claims specified on the IUA that (1) items designed using the systematically developed template elicit the same reasoning strategy, and (2) items may be solved using numeric relational reasoning. Similar to the methods used by Chesney et al. (2013), students' eye movements and fixations could be analyzed to determine if the students look back and forth between the expressions, which may indicate numeric relational reasoning, or if students primarily read the items from left to right, which does not indicate numeric relational reasoning. Combining these data with the cognitive interview data would present strong evidence about students' response processes with these items.

Limitations

It is worth noting that there are a few limitations not included in the preceding discussion. The student sample used in Study 2 was small, primarily white, and high-achieving. This sample does not represent a typical student population. Additionally, the size of the sample impacts the stability of the estimates. Although most items had acceptable point measure correlations and infit and outfit estimates, the results of this study serve as preliminary evidence, and additional data are needed. Studies with larger samples of students from a variety of backgrounds should be conducted and would provide additional evidence to substantiate the claims on the IUA.

Additional studies should also be conducted to systematically investigate students' response processes by property and item type. Interesting observations can be made using the data presented in this chapter, but because only a small number of items were selected per property or item type, these results cannot be generalized. However, the results help form hypotheses that can be investigated in future studies, particularly by conducting interviews and utilizing eye-tracking research methods.

Furthermore, evaluating the validity of the interpretations made from the ARPM system goes beyond just focusing on the two claims identified in this chapter. As previously mentioned, additional claims and assumptions should be added to the IUA in order to fully examine the validity of the interpretations made using the ARPM data and subsequent uses.

Conclusion

The purpose of this chapter was to describe the validity evidence gathered to examine the trustworthiness and meaningfulness of two claims made about the items on the ARPM system: (1) items designed using a systematically developed template elicit the same reasoning strategy, and (2) items may be solved using numeric relational reasoning. The evidence presented from two studies provides preliminary results indicating that the ARPM items designed using the template may elicit the same reasoning strategy and may be solved using numeric relational reasoning. However, because these studies have limitations, additional studies, using interview and eye-tracking techniques, should be conducted to strengthen the evidence about students' response processes and continue to support the evidence on the IUA for these claims.

References

American Educational Research Association, American Psychological Association, & National Council on Measurement in Education. (2014). *Standards for educational and psychological testing.* Washington, DC: American Educational Research Association.

Baker, F. (2001). *The basics of item response theory.* College Park, MD: ERIC Clearinghouse on Assessment and Evaluation, University of Maryland.

Behr, M., Erlwanger, S., & Nichols, E. (1980). How children view the equals sign. *Mathematics Teaching, 92,* 13–15.

Carpenter, T. P., Franke, M. L., & Levi, L. (2003). *Thinking mathematically: Integrating arithmetic and algebra in elementary school*. Portsmouth, NH: Heinemann.

Chesney, D. L., McNeil, N. M., Brockmole, J. R., & Kelley, K. (2013). An eye for relations: Eye-tracking indicates long-term negative effects of operational thinking on understanding math equivalence. *Memory & Cognition, 41*, 1079–1095.

Embretson, S., & Gorin, J. (2001). Improving construct validity with cognitive psychology principles. *Journal of Educational Measurement, 38*(4), 343–368. doi:10.1111/j.1745-3984.2001.tb01131.x

Ercikan, K., & Pellegrino, J. (2017). *Validation of score meaning for the next generation of assessments: The use of response processes*. New York, NY: Routledge.

Ericsson, K. A., & Simon, H. A. (1993). *Protocol analysis: Verbal reports as data*. Cambridge, MA: MIT Press.

Falkner, K. P., Levi, L., & Carpenter, T. P. (1999). Children's understanding of equality: A foundation for algebra. *Teaching Children Mathematics, 6*(4), 232–236.

Farrington-Flint, L., Canobi, K. H., Wood, C., & Faulkner, D. (2007). The role of relational reasoning in children's addition concepts. *British Journal of Developmental Psychology, 25*, 227–246. doi:10.1348/026151006X108406

Geary, D. C., Boykin, A. W., Embretson, S., Reyna, V., Siegler, R., Berch, D. B., & Graban, J. (2008). *Chapter 4: Report of the task group on learning processes*. Retrieved from http://www.ed.gov/about/bdscomm/list/mathpanel/report/learning-processes.pdf

Gersten, R., Beckmann, S., Clarke, B., Foegen, A., Marsh, L., Star, J. R., & Witzel, B. (2009). *Assisting students struggling with mathematics: Response to Intervention (RtI) for elementary and middle schools* (NCEE 2009–4060). Washington, DC: National Center for Education Evaluation and Regional Assistance, Institute of Education Sciences, US Department of Education.

Hatfield, C., Perry, L., Hayata, C. A., Geller, J., Barasch, B., & Ketterlin-Geller, L. (2016). *Imagination Station (Istation): Algebra readiness progress monitoring system development for grades 6–8* (Tech. Rep. No. 16–05). Dallas, TX: Southern Methodist University, Research in Mathematics Education.

Hubley, A. M., & Zumbo, B. D. (2017). Response processes in the context of validity: Setting the stage. In A. M. Hubley & B. D. Zumbo (Eds.), *Understanding and investigating response processes in validation research* (pp. 1–12). Cham, Switzerland: Springer.

Just, M. A., & Carpenter, P. A. (1980). A theory of reading: From eye fixations to comprehension. *Psychological Review, 87*, 329–354.

Kane, M. T. (2006). Validation. In R. Brennan (Ed.), *Educational measurement* (4th ed., pp. 17–64). Westport, CT: American Council on Education and Praeger.

Kane, M. T. (2013). Validating the interpretations and uses of test scores. *Journal of Educational Measurement, 50*(1), 1–73. doi:10.1111/jedm.12000

Ketterlin-Geller, L. R., & Chard, D. J. (2011). Algebra readiness for students with learning difficulties in grades 4–8: Support through the study of number. *Australian Journal of Learning Disabilities, 16*(1), 65–78.

Ketterlin-Geller, L. R., Gifford, D. B., & Perry, L. (2015). Measuring middle school students' algebra readiness: Examining validity evidence for three experimental measures. *Assessment for Effective Intervention, 41*(1), 28–40. doi:10.1177/1534508415586545

Lai, M., Tsai, M., Yang, F., Hsu, C., Liu, T., Lee, S. W., . . ., Tsai, C. (2013). A review of using eye-tracking technology in exploring learning from 2000 to 2012. *Educational Research Review, 10*, 90–115.

Leighton, J. P., Tang, W., & Guo, Q. (2017). Response processes and validity evidence: Controlling for emotions in think aloud interviews. In A. M. Hubley & B. D. Zumbo (Eds.), *Understanding and investigating response processes in validation research* (pp. 137–158). Cham, Switzerland: Springer.

Li, Z., Banerjee, J., & Zumbo, B. D. (2017). Response time data as validity evidence: Has it lived up to its promise and, if not, what would it take to do so. In A. M. Hubley & B. D. Zumbo (Eds.), *Understanding and investigating response processes in validation research* (pp. 159–178). Cham, Switzerland: Springer.

Linacre, J. M. (1994). Sample size and item calibration stability. *Rasch Measurement Transactions, 7*(4), 328.

Linacre, J. M. (2002). What do infit and outfit, mean-square and standardized mean? *Rasch Measurement Transactions, 16*(2), 878.

Linacre, J. M. (2012). *Winsteps Rasch measurement computer program.* Beaverton, OR: Winsteps.com.

Lord, F. M. (1980). *Applications of item response theory to practical testing problems.* Hillsdale, NJ: Erlbaum.

McNeil, N. M, & Alibali, M. W. (2005a). Why won't you change your mind? Knowledge of operational patterns hinders learning and performance on equations. *Child Development, 76*(4), 883–899. doi:10.1111/j.1467-8624.2005.00884.x

McNeil, N. M., & Alibali, M. W. (2005b). Knowledge change as a function of mathematics experience: All contexts are not created equal. *Journal of Cognition and Development, 6*(2), 285–306. doi:10.1207/s15327647jcd0602_6

Miller, K., Chepp, V., Wilson, S., & Padilla, J. L. (2014). *Cognitive interviewing methodology.* Hoboken, NJ: John Wiley.

Molina, M., & Ambrose, R. C. (2006). Fostering relational thinking while negotiating the meaning of the equals sign. *Teaching Children Mathematics, 13*(2), 111–117.

National Mathematics Advisory Panel. (2008). *Foundations for success: The final report of the national mathematics advisory panel.* Washington, DC: US Department of Education.

Oranje, A., Gorin, J., Jia, Y., & Kerr, D. (2017). Collecting, analysing, and interpreting response time, eye tracking and log data. In K. Ercikan & J. W. Pellegrino (Eds.), *Validation of score meaning for the next generation of assessments* (pp. 39–51). New York, NY: Routledge.

Padilla, J.-L., & Leighton, J. P. (2017). Cognitive interviewing and think aloud methods. In A. M. Hubley & B. D. Zumbo (Eds.), *Understanding and investigating response processes in validation research* (pp. 211–228). Cham, Switzerland: Springer.

Powell, S. (2012). Equations and the equal sign in elementary mathematics textbooks. *The Elementary School Journal, 112*(4), 627–648.

Rayner, K. (1998). Eye movement in reading and information processing: 20 years of research. *Psychological Bulletin, 124*(3), 372–422.

Rios, J., & Wells, C. (2014). Validity evidence based on internal structure. *Psicothema, 26*(1), 108–116. doi:10.7334/psicothema2013.260

Stephens, A., Knuth, E. J., Blanton, M. L., Isler, I., Gardiner, A. M., & Marum, T. (2013). Equation structure and the meaning of the equal sign: The impact of task selection in eliciting elementary students' understandings. *The Journal of Mathematical Behavior, 32*, 173–182. doi:10.1016/j.jmathb.2013.02.001

Wu, H. (2001). How to prepare students for Algebra. *American Educator, 25*(2), 1–7.

4 A Validation Approach to Middle-Grades Learning Trajectories Within a Digital Learning System Applied to the "Measuring Characteristics of Circles"

Jere Confrey and Emily Toutkoushian

Introduction

Learning trajectories (LTs) are discussed widely as a tool in mathematics education for a variety of purposes including designing curriculum, helping teachers to elicit student thinking, strengthening teacher knowledge, and assessing student progress. With these multiple uses in mind, our team has designed and implemented a digital learning system (DLS) (Confrey, 2015) built around the concept of learning trajectories for use with middle school mathematics students in grades 6–8. The DLS, called Math-Mapper 6–8 (MM6–8), consists of a learning map organized hierarchically around big ideas, a diagnostic assessment and reporting system, and a variety of types of scaffolding for digital curricular resources. Assessments designed to generate feedback on student learning in real time have been recently labeled as classroom assessment (National Council on Measurement in Education [NCME], 2017) and share some key qualities of formative assessment and diagnostic measures in that they are designed to give precise, immediately relevant, and actionable feedback. Classroom assessment differs from most informal formative assessments because psychometric approaches are applied to validate the measures and support their connection to other components of the assessment system, such as interim tests and end-of-year tests (Shepard, Penuel, & Pellegrino, 2018). In this chapter, we describe how we measure students' progress along a learning trajectory using classroom assessments and digitally return those data to students and teachers in real time to improve instruction. Then we describe our validation framework for learning trajectory measurement and the results of a validation study of one cluster called *Measuring Characteristics of Circles*.

The concept of validation began with the articulation of various types of validity (content, construct, criterion, and consequential). It evolved to a set of claims about validity including: (a) one does not validate a test but rather validates how that test is used for a particular purpose; (b) validation is "an integrated evaluative judgment of the degree to which empirical evidence and theoretical rationales support the adequacy and appropriateness of inferences and actions based on test scores" (Messick, 1989, p. 13); and (c) validation is an ongoing process. Fundamentally, because one is striving to measure a psychological construct that cannot be measured directly, validation requires an interpretive argument that provides the necessary warrants for the propositions and claims of those arguments (Haertel & Lorie, 2004; Kane, 2006; Mislevy, Steinberg, & Almond, 2003). As emphasized in the assessment triangle from *Knowing What Students Know* (National Research Council, 2001), validation requires one to connect the *cognitive* construct under investigation, the means to *observe* that construct and gather evidence, and the *interpretation* of that evidence.

Our contribution to the field rests in treating validation as an iterative process in a DLS with a relatively continuous flow of information from users and digital affordances supporting change and adjustment. Our user models are refined in real time as we conduct observation-based design studies and consult with teachers and administrators regularly. Hence, validation is dynamic. The fundamental challenge is to continuously improve the quality of the measurement in the system.

We discuss learning trajectories and situate them in our DLS describing our diagnostic assessment and reporting system. Our overall validation framework is introduced, followed by a particular validation study. We apply our measurement model and conduct and interpret related analyses to create a validation argument aligned to the framework. We conclude by outlining how the study and its analysis support the revision and improvement of LTs over time.

Learning Trajectories

Learning trajectories are research-based descriptions of the landmarks and obstacles that students often encounter as they move from naïve to sophisticated thinking about a target construct (Clements & Sarama, 2004; Confrey, Maloney, & Nguyen, 2014; Lehrer & Schauble, 2015). They consist of a target construct and a description of the ideas students bring to instruction and behaviors they exhibit, indicating different levels of sophistication as they proceed through instruction. This allows us to build assessments that can be used with various curricula and to assess their effects on progress along LTs.

Our MM6–8 team describes an LT as a set of progress levels (Wilson, 2005). Levels include descriptions of the representations, strategies, cases, properties, and generalizations exhibited by students that provide evidence of more sophisticated reasoning. We emphasize that learning trajectories are not stage theories (Confrey, Maloney, Nguyen, & Rupp, 2014; Lehrer & Schauble, 2015). A student does not necessarily have to master all the lower stages to learn a higher level. Instead we argue that movement in an LT is more akin to movement up a climbing wall replete with handholds and obstacles (Confrey et al., 2019). It is expected that students start in different places and pursue various paths up the wall and that the handholds and obstacles are predictable. We see predictions of student paths as probabilistic.

The learning trajectories in MM6–8 sit within the hierarchical organization of the map. The primary organizer of the content for middle school mathematics is a set of nine *big ideas*. The big idea central to the validation study presented herein is "Measure, compose, and scale perimeter, area and volume" within the field of geometry. In MM6–8, a big idea is composed of relational learning clusters, each of which is a set of constructs that should be taught in tandem because they address closely related ideas. The overall map consists of 24 clusters. Each cluster comprises constructs, and for each construct an LT is articulated. Across the MM6–8 map, there are 63 constructs and related LTs. For this validation study, we focus on the seventh grade cluster: *Measuring Characteristics of Circles* with its two constructs, *Pi and Circumference* and *Area of Circles* and their LTs. The LT for *Pi and Circumference* is:

L1: Recognizes that a family of regular *n*-gons are similar, leading to a scale factor of *n* relating the sides and the perimeters

L2: Estimates pi by comparing the lengths of circumferences to the diameters as ratios

L3: Names the ratio of circumference to diameter as pi

L4: Knows pi is an infinite decimal with approximations of 3.14 or 22/7

L5: Solves problems involving radius or diameter and circumference by working with 3.14 or 22/7 as approximations for pi

L6: Solves problems involving radius or diameter and circumference by working with pi, as a symbol and a constant.

The LT for *Area of Circles* is:

L1: Estimates the area of a circle using different-sized square units

L2: Estimates the area of a circle by modeling it with one or more polygons

L3: Uses and justifies the formula for area of a circle

L4: Uses area and circumference formulas appropriately to solve problems in context.

Some levels of an LT have one or more misconceptions associated with the level. For example, at level 5 in the LT for pi and circumference is the misconception: "Scales radius by pi (π) to find circumference" (It should be *scales diameter or twice the radius*). Each construct is aligned to the

CCSS-M standards (CCSSI, 2010). Teachers are encouraged to focus their attention on the LT to drive instruction while addressing the standards.

A key role of the learning trajectory in MM6–8 is to understand that using LTs to build diagnostic assessments differs from typical domain sampling approaches to assessment (Briggs, 2017). Typical assessments use domain sampling, in which tests are constructed by identifying the subconstructs to be tested and then randomly choosing items from those subconstructs. By contrast, in LT assessment one uses the sequencing of levels to structure the test. The LT framework is simultaneously visible and explicit for both instruction and assessment purposes.

Diagnostic Assessments as Situated in Math-Mapper 6–8

The diagnostic assessments are designed to measure students' progress along LTs at the construct level, but by administering the assessments at the cluster level, we stress the coherence of the ideas in the cluster and avoid too much testing. Most teachers measure their students' progress along the LTs of a cluster about two-thirds of the way through a teacher's instructional unit, allowing sufficient time to make any changes to instruction. An assessment consists of 8–12 items, administered digitally and lasting 20–30 minutes. Multiple forms of a cluster-based assessment are spiraled within an individual class to ensure measurement of more LT levels across the class.

MM6–8 accommodates varied curricula, as test timing can be customized. Teachers assign tests either for a given date or within a specified time interval. Tests can be designated as tests, retests, or pre- or post-tests. Two or more items are written for each progress level of the LT. Item types include multiple choice (33%), multiple select (20%), numeric (28%), one-letter (18%), and open-ended (1%).

Having multiple forms increases the diversity of the items encountered by students and reviewed in classes. We implemented three design features to ensure that the forms were equivalent and the scores could be compared: blueprints to maintain consistent domain representation across forms, form assembly to minimize differences in mean item difficulty, and item response theory (IRT) scoring based on calibrated item pools. The blueprints were created to ensure that the number of items per test form were proportional to the number of levels within each construct. Common items were selected to be included across all forms and agree with the blueprint proportionally, so that the common items themselves make *mini-tests*, which unlike testlets are comparable in content representativeness to the full test forms and are considered together with the items in the full form (Kolen & Brennan, 2004). To minimize the difference in mean item difficulty across forms, we used OpenSolver optimization software within Microsoft Excel to minimize differences in the expected mean percent correct score while satisfying blueprint and time constraints (Mason, 2011). To further improve comparability of the scores, IRT scale scores are used rather than raw scores: we calibrated all field-tested items using the MIRT package in R (Chalmers, 2012) to place them onto the same theta scale. Theta scores from each form were then placed onto the same scale, rendering the scores comparable across forms and constructs.

Data Reporting

A completed assessment is digitally scored; the resulting outcome data are immediately returned to students and teachers for review and planning subsequent activities. Students receive scores as percent correct for each construct. A "learning ladder" for each trajectory lists the progress levels and shows (color-coded) the proportion correct for levels at which they were tested. An item matrix displays the credit received and allows them, if a response was incorrect, to revise and resubmit for more credit.

Teachers receive the outcome data in a class report that summarizes the class performance by level using representations called *heatmaps* (Figure 4.1). Heatmaps display results by construct for a cluster. The heatmap for each construct lists the levels of the LT vertically from lowest to highest. Along the horizontal axis, the student names are listed from lowest

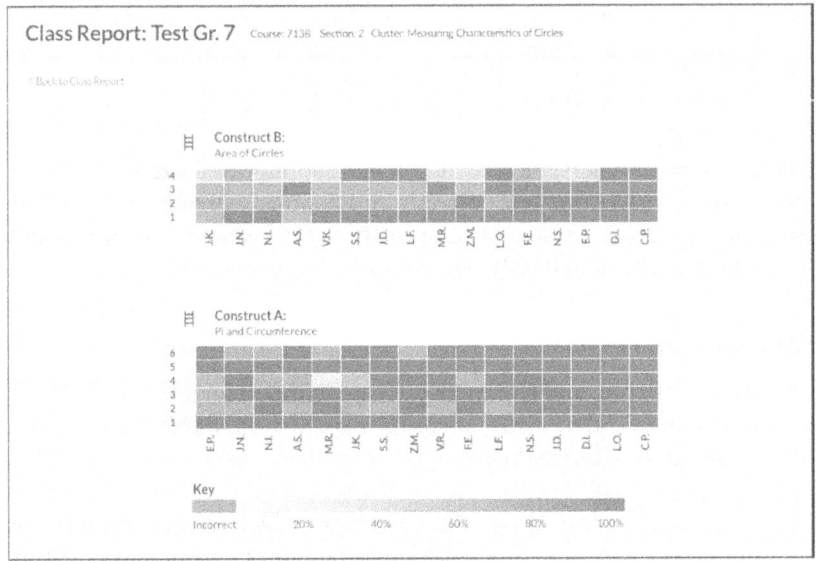

Figure 4.1 Class report: heatmaps for the two LTs of this chapter. Each cell represents a single student's score on an item measuring a single level. Vertical axis: LT progress levels, ordered from low (bottom) to high (top). Horizontal axis: students ordered by overall construct score, from low (left) to high (right).

to highest cluster theta scores. Each cell represents a student's score for a single item in a level and is color-coded from light gray (incorrect) to dark gray (correct). If students did not receive an item at that level on their form it would appear as white. A solid-colored row (without white) indicates a common item. The overall expectation is for students to perform better on the lower levels and less well on the upper ones. If this condition applies to a particular class's performance, the boundary between the light and dark squares creates a curve called a "Guttman curve"; as a class improves on a cluster, that curve moves to the left and upward.

Teachers use the overall shape to determine which levels and students need more attention and instructional treatment. Teachers can work with subgroups or the whole class in reviewing the data. A major goal of MM6–8 is to promote more learner-centered discussion of the ways students approached the problems, and how those problems relate to the levels and the overall understanding of the target construct. One aim of the design is that the teacher mediates this discussion using a variety of techniques that involve the selection of the items, the role of students in the review process, the use of feedback mechanisms and item analysis data, and the way the structure of the LT and map are leveraged in the discussion. Some teachers review items from lower to upper levels, while others focus only on the common items. Some invite students to explain their methods, while others tell students how to solve the problem, which is less consistent with the intention of the design.

There is considerable variation in the degree to which teachers use the structure of the LT in their discussions when they review an assessment. At one extreme, teachers do not refer to the LT at all, treating the data as if it were from a domain sampling–based assessment. This approach is typically oriented only at whether a student's item response was correct or incorrect and how to answer the item correctly. At the other extreme, teachers draw students' attention to the meaning of the level associated with the item, stressing that the item is only one instance of the idea at that level.

Validation Framework

Pellegrino, DiBello, and Goldman (2016) identified three components for constructing a validation argument from assessments designed to inform instructional decision making:

1. Cognitive, "address[ing] the extent to which an assessment taps important forms of domain knowledge and skills in ways that are not confounded with other aspects of cognition" (p. 4);
2. Instructional, referring to the alignment of the assessment to the curriculum, students' opportunity to learn, and to "[use of the] assessment outcome information for teachers and students as a guide to instruction and learning" (p. 4);

3. Inferential, referring to the "extent to which an assessment reliably and accurately yields model-based information about a student performance, especially for diagnostic purposes" (p. 4).

Confrey et al. (2019) adapted and elaborated on the Pellegrino et al. framework to create a validation framework tailored to MM6–8. The modifications of this approach for MM6–8, and applications of a principled assessment design perspective (Nichols et al., 2016), are described below.

Cognition

Learning trajectories are the fundamental cognitive component for MM6–8. Synthesis of research is required to lay the foundation for an LT. This is typically followed by a series of clinical interviews or teaching or design experiments designed to elicit and document different aspects of student thinking over time. Such studies utilize either a cross-sectional or a longitudinal design. Six criteria are applied to LT progress levels in our approach (adapted from Pellegrino et al.):

- LTs are linked to significant and worthy targets, related to the big ideas and standards;
- LTs exhibit clear and comprehensive coverage, describing commonly elicited or observed student behaviors;
- Progress levels are ordered by increasing sophistication, with grainsizes distinct enough to foster measurement;
- Multiple entry points are supported;
- LTs incorporate the probabilistic character of student progress;
- LTs facilitate upward movement via instruction.

The last two criteria require explanation. LT levels are listed in order of increasing sophistication, but not all students will experience the same degree of challenge at each level. Some may get stuck at a level, while others may appear to practically skip it because it seems so evident. A climbing wall has become, for us, a useful metaphor to connote students' possible multiple paths within a predictable diverse set of obstacles and landmarks and varying difficulty of "routes" to the target understanding. Progress on such a wall can be quantified as height above the ground just as progress levels are sequenced in reference to an average difficulty of solving related problems. Spacing of handholds and particular types of obstacles are also factors that influence progress up a wall. Similarly, the demand of the cognitive activity at different levels influences difficulty along an LT. However, we do use a ladder to compactly and efficiently describe sequences of levels by difficulty, even though it must be noted that the ladder representation can lead to the misconception that there is, or should be, one single "best" or "most efficient" path. Our theoretical position is that there is a need for teachers to support multiple paths (wall metaphor) while anticipating likely patterns of student behavior (ladder/level metaphor). Implications of this structure become evident in psychometric modeling of the learning trajectories and in expectations of what is in the data.

The final criterion above acknowledges the role of instruction in progressing along LTs. In our estimation, progress along LTs also requires that students engage in activities directly aiming at teaching the ideas, whether through whole class instruction, peer-to-peer work, or individually engaging with curricular materials.

Instruction/Implementation

The goal of classroom assessment is to improve students' learning. A validation argument deriving from this framework must take into consideration how the assessment is implemented.

> In addition to challenging the sequence and specification of levels, another aspect of validity that must be examined with respect to a learning progression is its instructional efficacy, or the degree to which it is used successfully in the classroom.
>
> (Graf & van Rijn, 2016, p. 167)

The MM6–8 diagnostic assessments are administered during the time period the curricular material is taught, with the goal of providing valid, reliable, and timely feedback to the students and teachers regarding what students have learned in terms of progress along the LTs. In addition, examination and interpretation of the data can be viewed as a means for the teachers, as members of a professional learning community (PLC), to examine, analyze, discuss, and interpret the data over both short and long time frames. We specify four criteria for the implementation/instruction component of validation:

- LT-based assessments are aligned coherently with curriculum scope and sequences, and with instruction;
- Students and teachers conduct review of class and student reports;
- Teachers orchestrate post-assessment follow-up, based on prior assessment data;
- Groups of teachers adjust and revise subsequent curricular approaches based on assessment outcomes.

Inference

This component, "concerned with the extent to which an assessment reliably and accurately yields model-based information about student performance, especially for diagnostic purposes" (Pellegrino et al., 2016, p. 62) addresses linkages among inference, evidence, and data. Analytic methods from measurement and statistical inference are applied to analyze data from students' task performances, to discern the degree to which they "reliably align with one or more underlying conceptual measurement models that are appropriate to the intended interpretive use" (ibid., p. 62).

The majority of validation studies of LTs in mathematics or statistics have been conducted following the Berkeley Evaluation and Assessment Research (BEAR) method, which relies on four principles: a developmental perspective, a match between instruction and assessment, management by teachers, and evidence of high-quality assessment (Wilson, 2005). Many mathematics education studies that apply the BEAR approach (and others) focus on establishing that the ordinality of the levels is defensible using a Rasch model (Carney & Smith, in press; Clements, Sarama, & Liu, 2008; Confrey et al., 2014; Lehrer, Kim, Ayers, & Wilson, 2014; Pham, Bauer, Wylie, & Wells, in press; Wilmot, Schoenfeld, Wilson, Champney, & Zahner, 2011). These approaches assume that underlying the qualitative shifts between levels there is a continuous dimension that, if unidimensional, can be modeled as *difficulty*. Others take the view that the levels are discrete, and deploy cognitive diagnostic models (CDM) or latent class analysis models (Lai, Kobrin, DiCerbo, & Holland, 2017). Pham et al. (in press) have applied both a Rasch and CDM analysis to LTs on linear functions and proportional reasoning, and (allowing for minor shifts in ordering) found both models viable for explaining their data.

The MM6–8 system is complex and impossible to validate as a whole. Our approach has instead been to create a matrix that incorporates the three components (cognition, instruction/implementation, and inference) with six purposes for using the application. Those six purposes have been articulated as:

1. Inform teachers of class progress by LT
2. Elicit and leverage diverse levels of student thinking
3. Increase students' awareness of their own learning and growth
4. Strengthen teachers' content and pedagogical knowledge
5. Connect instructionally proximal data to more distal forms of assessment (interim and high stakes)
6. Improve student's knowledge of target constructs and big ideas from the map.

We identify one or more purposes as the focus for each particular validation study and attend to elements of the three components in order to form our validation argument.

A Validation Study

Having defined our foundational construct (learning trajectories), summarized the relevant features of MM6–8, and provided an overview of our validation framework, we now document a study of the validation argument for one cluster, *Measuring Characteristics of Circles*. Our validation process is carried out collaboratively among three teams of experts: measurement, learning sciences, and practitioners (Confrey et al., 2019). The primary focus of the study is purpose 1, informing teachers of class progress by LT. The question under examination is:

> To what degree can we provide a teacher with accurate and valid information on the class's progress along a learning trajectory for the purpose of adjusting and improving subsequent instruction to meet the students' needs?

The data for the study were collected during MM6–8 field-testing (2016–2018), conducted in two districts, at three partner schools with 1:1 computing. The first district, listed as low-performing at the state level, implemented 1:1 computing in 2015. District 1 serves 977 students (27% African American, 1% Asian, 10% Hispanic and mixed, 53% white, and 56.9% eligible for free/reduced lunch). One middle school has participated and has logged 9,000+ tests administered by 19 teachers. District 2, which implemented 1:1 in 2013, serves mostly upper-middle-class children and has been recognized as high-performing. It has two middle schools and serves 1,163 students (4% African American, 9% Asian, 8% Hispanic or mixed, 79% white, and 9.9% eligible for free/reduced lunch). The two schools have logged 21,000+ tests administered by 33 participating teachers.

Claims Regarding the System Design and Use

Five specific claims about the system design with respect to its measurement theory, related to the classroom use of heatmaps by teachers and students, are identified to make up our validation argument:

> *Claim 1.* The construct subscores for a cluster will be highly correlated, reflecting mutual dependencies implied by the structure of the cluster.
> *Claim 2.* Overall, constructs will demonstrate a strong positive relationship between the empirical difficulty of items and LT levels, indicating a trend of increasing sophistication among levels.
> *Claim 3.* Empirical item difficulties will vary within an LT level in ways closely associated with the meaning of the level.
> *Claim 4.* Ordering the levels, and the items by difficulty within the levels (heatmaps), will reveal relative strength of students' performances across items and levels and across the class.
> *Claim 5.* Teachers target instruction based on the data on students' performance by level.

Together, these claims provide a chain of reasoning from observed item scores to intended interpretations: if the internal structure of the item pool sufficiently reflects the constructs and LTs of the learning map, the heatmaps will be accurate representations of student knowledge and understanding (claims 1–3), and the final step in the argument will depend on teacher's ability to read and make use of the heatmap with their students (claims 4–5).

This specific validation study will focus on the first three claims in the validation argument which in turn form the foundation that supports claims 4 and 5. Claims 1 through 3 are necessary, but not sufficient conditions to establish the validity of using MM6–8 for the purposes outlined above. The scope of this chapter excludes claims 4 and 5, to be addressed in a future analysis.

Role and Assumptions of the Measurement Model

The LT-based assessments in MM are new measures within a large, complex system of relationships among LTs. In this validation process, measurement approaches are used as research tools to support analysis and interpretation of the assessment data in relation to the underlying LTs. The goal is to identify the simplest measurement model that will facilitate identification of (a) items that do not appropriately measure their progress levels, (b) progress levels that are either poorly defined or out of order, and (c) explanations for (a) and (b). This effort has been separated into sequential steps to construct a principled and manageable process.

Progress levels (1 through n) for an LT are designed to increase in sophistication, with upper progress levels and their items for a given construct being more challenging for most students to learn than lower progress levels and their items. We anticipate that, on average across many students, the difficulty of the items measuring progress levels will monotonically increase with levels.

We do not predict, nor necessarily desire, that all items at one level of an LT are less difficult than all items at the next level (i.e., not a stage theory model). There are at least two reasons to allow for substantial variation within a level, and overlap with the variation at other levels: (1) in mathematics, cases associated with a specific level can legitimately vary in difficulty based on many factors, and (2) students' performance on these items can be influenced by instructional factors, including those connected to opportunity to learn. These considerations led us to distinguish three categories of variability affecting the difficulty of items:

- Variation in difficulty among items that reside in a single progress level within an LT: *intra-level variation* ("Intra-LV"); such items may address different facets of reasoning at a single level;
- Variation in item difficulty *between* levels within an LT: *inter-level variation* ("Inter-LV");

- Variation associated with bias and noise: *construct-irrelevant variation* ("Irrel-V"). Sources may include ambiguous wording, a poor or misleading representation, or students' lack of familiarity with a problem context.

With these distinctions in mind, we state the overall question for the validation study:

> To what extent does the underlying correlational and empirical difficulty structure of the item scores, construct scores, and cluster scores support inferences concerning students' positions in an LT, and how can the structure systematically be improved?

We answer this general validation question via responses to four individual questions, each related to one of the three claims (1–3).

Question 1. Are the Construct Scores Independent or Mutually Dependent?

Procedure for Q1: Item Response Theory Analyses

To investigate the correlational structure of the test data, we compared the fit of two confirmatory item response theory (IRT) models. First, we estimated a unidimensional model (Model 1), which is equivalent to a model that assumes the correlation of each construct score to be equal to 1 (Hatcher, 1994). Second, we fit a two-dimensional model (Model 2) that assumed a simple structure, where each item aligns to the construct it is supposed to measure. In this model, we estimated the correlations of the latent construct scores. If Model 1 fits the data better than Model 2, we take this as evidence that the construct scores are not independent but are mutually dependent. We expect that the construct scores will be mutually dependent because the constructs are nested together within the same cluster.

MIRT software (Chalmers, 2012) in R was used to perform single-group and multigroup IRT calibrations comparing the fit of various models to determine the best-fitting model for the mixture of dichotomous and polytomous items using the following fit statistics: Bayesian information criterion (BIC), sample-size adjusted BIC (SBIC), Akaike's fit index (AIC), and corrected AIC (AICc). We also applied the chi-square difference test when comparing nested models and used the Metropolis-Hastings Robbins-Monro algorithm (MHRM) for all calibrations (Cai, 2010).

Results for Q1

The correlational structure of the 20 items in the cluster's item pool was examined, both in terms of how well the scores covaried with themselves via marginal reliabilities, and in terms of how the two construct scores correlated with each other. The model-based marginal reliability estimate for the cluster score from Model 1 was 0.56; the reliabilities for the construct scores from Model 2 were 0.54 and 0.51. Because the reliabilities of the construct scores were below the minimum reliability of 0.60 suggested by Hatcher (1994), we examined the data set to determine if there were any sources of data that may have contributed to the low marginal reliability estimate. One identified data source was small hybrid forms requested during our first year of field-testing, which were composed of samples of items from different clusters ranging from one item per cluster to five items. The smaller hybrid forms with less than four items were determined to be different from the rest of the records and removed from the data set, because the average score on items was about 11% lower than the rest of the records, and the population of students taking these forms was from only one of our two districts.

Once those smaller hybrid forms were removed from the data set, the reliability of the scores improved closer to the desired threshold of 0.60. The Model 1 marginal reliability was 0.59, and the Model 2 reliabilities became 0.57 and 0.54. Further questions in this analysis consider potential items that may not fit their LT level, so these marginal reliabilities were deemed appropriate to proceed.

In order to begin investigating the independence of the two constructs, the correlation of the two latent construct scores (0.90) in Model 2 was considered. To determine if a correlation of two factors is statistically significant, Anderson and Gerbing (1988) recommend calculating the upper confidence interval of the correlation to see if the value includes 1. Because the 95% confidence interval (0.821, 1.004) contains the value 1, we concluded that the correlation is equivalent to 1. This evidence suggests that the two constructs are mutually dependent and unidimensional.

Comparing the global fit indices for two models provides further illustration of unidimensionality: Model 1 clearly fits better than Model 2 (Table 4.1), for not only are the AIC, AICc, SBIC, and BIC fit indices smaller for Model 1, but the chi-square difference test of the two models is rejected ($p > 0.05$). Because the correlation of the construct scores was statistically indistinguishable from 1, and the global fit statistics indicated the unidimensional model fit better than the multidimensional model, we concluded that Model 1 was supported over Model 2, providing strong evidence that the construct scores are mutually dependent, not independent.

After determining the data to be unidimensional for the two constructs, we compared the fit of two competing unidimensional IRT models: the Rasch model and two-parameter logistic (2PL) model. The fit statistics provided mixed results (AIC, AICc, SBIC favoring the 2PL model and BIC and log likelihood favoring the Rasch model), but comparing the χ^2 change statistic (de Ayala, 2009), which quantifies the difference in variance explained by two models, reveals a mere 1.09% improvement in fit

Table 4.1 Comparison of Fit Statistics of Two Confirmatory IRT Models

Model	AIC	AICc	SBIC	BIC	logLik	X² diff	df	p
1: equal slopes 1-D IRT	8,928.06	8,931.04	8,980.87	9,088.84	−4,430.03	NA	NA	NA
2: equal slopes within each LT, 2-D MIRT	8,929.42	8,933.76	8,983.33	9,097.65	−4,430.27	−0.48	2	0.12

of the 2PL over the Rasch model. The small practical differences between the models was also reflected in the extremely high correlation of the theta scores of the two models ($r = 0.98$). Given the negligible improvement in model fit and high correlation between the models, we selected the simpler Rasch model for the rest of the analysis.

Discussion of Q1

The results supported the decision to consider the two constructs, *Pi and Circumference* and *Area of Circles*, as unidimensional rather than independent of each other. This was supported both by considering the correlation of the two constructs in the multidimensional model and comparing the global fit statistics between a unidimensional and a multidimensional model.

Question 2: To What Extent Does the Pattern of Item Difficulties Within and Across LTs for a Cluster Agree With the Pattern of Content Sophistication Within and Across LTs?

Procedure for Q2

This section examines the structure of the item pool through the lens of empirical item difficulty, where difficulty is defined by the Rasch model *b* parameter from Model 1. Because the Rasch model provides *b* parameters for each of the possible points in polytomous items, we chose to use the *b* parameter associated with getting all parts of an item correct (full credit) to aid in the comparability to the dichotomous items which are scored that way. The Guttman structure, alluded to above in the context of heatmap reports *by construct*, is closely associated with a strong positive relationship between LT level and item difficulty. Therefore, if a strong positive relationship is present, the Guttman structure is also present. To answer Q3, we utilized linear regression to examine the relationship between item difficulty (*b*) and LT level, expressed as discrete numbers (1, 2, 3, . . .) for each construct. We interpreted models with a positive slope and a high R^2 as evidence that the pattern of content sophistication in the LTs is indeed reflected in the pattern of empirical item difficulties. Additionally, in order to determine if mis-fitting items were distorting the regression, after running a baseline model we sequentially removed a single item based on its absolute residual value, reran the regression, and compared the new R^2, residual sum of squares (RSS), and slope to the previous regression. We wanted to maximize the R^2 value above 0.60 while also retaining as many items as possible. Finally, we examined the Spearman rank correlation of LT level and the item difficulty parameter b for the remaining subset of items. We interpreted strong positive correlations as further evidence that the pattern of content sophistication in the LTs are indeed reflected in the pattern of empirical item difficulties.

Results for Q2

Figure 4.2 and Table 4.2 show the results of the sequential regression models generated for each of the constructs. The baseline model for *Pi and Circumference*, which included all of the items, had a slope only minimally above zero (0.09) and an R^2 value of 0.01 and negative adjusted R^2, suggesting that the relationship between LT level and item difficulty for this model was barely positive and only 1% of the variance in LT level was explained by item difficulty. To improve the fit of the regression line, the absolute residual value for all of the items was considered; the item with the largest value was removed (Item 349, $n = 554$). The regression run with the remaining 18 items was an improvement from the baseline model: the slope increased to 0.23, the sum of squares decreased, and the R^2 increased to 0.10. Following the same pattern, two subsequent regressions were run, first removing Item 453 ($n = 676$) and then Item 1021 ($n = 211$) with the sum of squares decreasing and R^2 increasing each time, but still failing to reach the desired 0.60 threshold for R^2 (Item 453: 0.33, Item 1021: 0.38). We considered removing one additional item based on its absolute residual value (Item 526, $n = 283$). The final regression without Item 526 displayed marked improvement over the previous model, with slope increasing, sum of squares decreasing, R^2 increasing over 0.50 to 0.55, and the adjusted R^2 increasing to 0.49, providing evidence that this item should remain excluded from the final model.

The baseline regression for the *Areas of Circles* construct exhibited a much better fit than that for *Pi and Circumference*, with a positive slope of 0.76 and R^2 of 0.46. This baseline model already exhibited adequate fit, suggesting a strong positive relationship between LT level and item difficulty, and the R^2 was near the desired threshold, nonetheless removed one item and compared the results to determine if the item should remain eliminated. When Item 269 ($n = 310$; with largest residual value) was removed, the model improved significantly: the slope increased to 0.87, the sum of squares decreased, and the adjusted R^2 increased to 0.90, suggesting that in this model 90% of item difficulty was explained by LT

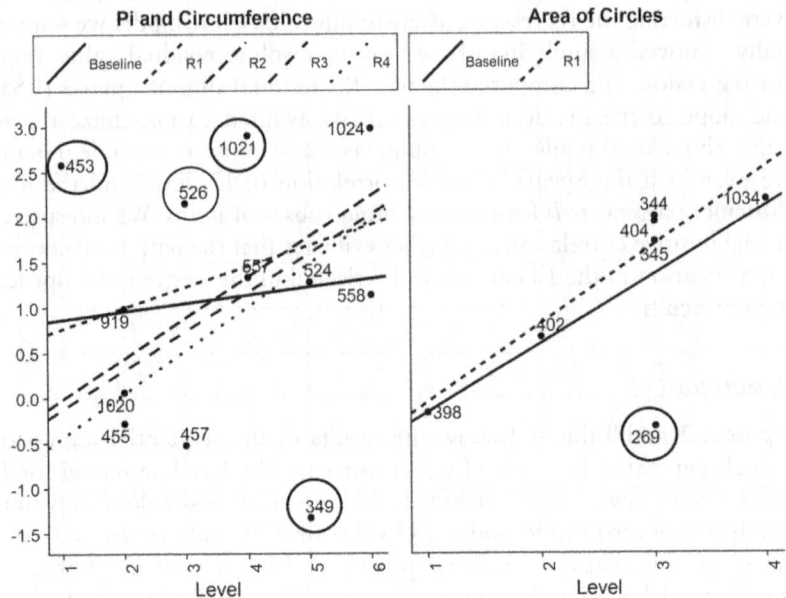

Figure 4.2 Linear regression of LT level (*x*-axis) predicting item difficulty (*b*, *y*-axis). Patterned regression lines for each of the regressions run sequentially after removing one item based on absolute residual value. Circled items (pattern matches line) were removed and remained out of the model for subsequent regressions.

Table 4.2 Regression Equations, R^2, and Sum of Squares for Sequential Regression Models

Construct	Model	Slope	Intercept	SS	R^2	Adj. R^2
Pi and Circumference	Baseline (All Items)	0.09	0.78	30.55	0.01	−0.09
	R1: Without Item 349	0.23	0.54	13.76	0.10	0.0
	R2: Without Items 349 and 453	0.45	−0.46	9.01	0.33	0.25
	R3: Without Items 349, 453, and 1021	0.43	−0.55	6.28	0.39	0.30
	R4: Without Items 349, 453, 1021, and 526	0.48	−0.91	3.95	0.55	0.49
Area of Circles	Baseline	0.76	−0.92	3.72	0.46	0.35
	R1: Without Item 269	0.87	−0.92	0.33	0.92	0.90

level. We determined this was enough evidence to suggest that Item 269 was potentially problematic and leave it removed from the model.

Finally, to provide further evidence about the relationship between LT level and item difficulty, the Spearman rank correlation coefficient was computed for the final model in each construct. Both indicated a strong positive correlation, significant at the $p < 0.10$ level: 0.67 for *Pi and Circumference* ($p = 0.07$) and 0.77 for *Areas of Circles* ($p = 0.07$).

Discussion of Q2

After removing five items that were sequentially identified as potentially non-conforming, based on their fit to a regression line modeling the desired positive relationship between item difficulty and ordered LT levels, the model of remaining items for each construct provided evidence of the increasing content sophistication of LT levels. The R^2 and adjusted R^2 values suggest how much of the variance in item difficulty can be explained by LT level in the regression model with the adjusted R^2 correcting for the number of predictors. Compared to the first model with all of the items included and only 1% of the variance in item difficulty explained by LT level, the final model for *Pi and Circumference* was able to account for 55% of the variance in item difficulty. This improvement suggests that the general pattern of difficulties does follow the desired pattern of LT levels. The *Area of Circles* construct appeared to have an even better fit of item difficulties to LT levels, with only one item being removed, it had a much higher R^2 value than *Pi and Circumference*. The high and positive values of the Spearman rank correlation coefficient for both of the constructs provided additional evidence supporting the agreement in the patterns of items and LT level.

Question 3: Which Items at Which Levels Show Evidence of Non-conformance to the Pattern of Item Difficulties Within and Across LTs, What Are the Likely Causes of Such Non-conformance, and What Should Be Done With Such Items?

Procedure for Q3

The following procedure is used to examine each item provisionally designated as non-conforming:

1. The item is reviewed by the learning sciences team, who examine its structure and empirical response behavior relative to its position in the LT, and suggests reason(s) for the item appearing not to conform;
2. Categories of variation that are relevant to each item are assigned: construct-irrelevant variation (Irrel-V), Intra-level (Intra-LV), or Inter-level (Inter-LV) variation;

3. Recommendation made for action(s) based on the analysis of variation among the following choices:

   a. Retire the item
   b. Revise item to remove irrelevant variation
   c. Revise item to adjust difficulty within level
   d. Maintain item as is, and post an alert indicating that this item is known to be unusually hard or easy
   e. Edit the level's description to clarify range of or inclusion of items
   f. Move the item or set of items to another level
   g. Adjust order of levels.

Our standard procedure when revising items is to retain the original item, field-test the revised item, and compare the results. Action b addresses the issues associated with Irrel-V, actions c–e addresses Intra-LV and actions f–g address Inter-LV. Examination for Irrel-V sought factors such as language or readability, representational ambiguity, use of unfamiliar terminology or context, being solvable by test-taking tricks, or other construct-irrelevant distractions such as too many steps, too many distractors, or too much time required to solve it. Irrel-V can also be identified by psychometric flags unrelated to the regression screening, during testing. Such psychometric flags are triggered if the differential item functioning (DIF) statistics, readability, word count, duration, or model-based misfit exceeds a threshold. These are reviewed on an annual basis.

Unlike Irrel-V, which one seeks to eliminate, an appropriate degree of Intra-LV is desirable because the items should assess the different aspects of the meaning of the level. Mathematical knowledge often results from acts of generalization by seeing invariant relations among different instances of an idea. It can play an important role instructionally, in promoting important cognitive challenge and rich classroom discourse. Understanding more about the Intra-LV variation is one of our research goals. General Intra-LV factors identified to date include mathematical issues concerning the numeric values, directness of the question, familiarity or ease with the representation, availability of a calculator, or availability of additional visual support. The analysis is conducted by examining the description of the level, associated misconceptions, data from the item analysis, and other items at the level for comparison. If adjustments to the item are proposed, the original item can be kept or retired. If kept, the revised item becomes a clone, and performance of both the original and the clone can be compared.

If an item or all the items at a level are identified as showing Inter-LV, the item(s) is/are considered relative to other levels. If the item fits better there in substance and difficulty, the item(s) is/are moved. Movement to a different level requires that the stepwise regression be run again to check whether different items become potentially non-conforming.

Results for Q3

> **Item 349** (*Pi and Circumference* L5: "Solves problems with an approximation of pi"); difficulty parameter $b = -1.3$ (predicted: 1.49). It asked students to identify which feature of a circle could be associated with a given set of measurements and their units. On review, this item was recommended for retirement because it proved poorly aligned to the level.
>
> **Item 453** (*Pi and Circumference* L1: "Recognizes that a family of regular *n*-gons are similar, leading to a scale factor of *n* relating the sides and the perimeters"); difficulty parameter $b = 2.6$ (predicted: 0.53). It is a T/F item with five statements. Each statement in the item describes the perimeter (or circumference) of a shape (rectangle, equilateral triangle, circle, hexagon, or general triangle), compared it to a linear dimension of the shape (a side, chord, or diameter), and asked if the ratio was constant. The results were mixed across the different options. The highest percentage of students understood the ratio of the perimeter length (circumference) to side, chord, or diameter length would not be constant for any triangle (71%) or rectangle (63%), and would be constant for an equilateral triangle (64%) or regular hexagon (54%). Only 45% recognized a constant ratio for circles. The team argued that this suggested that this insight into π is likely not to be recognized by teachers. We also believed the all the examples were needed to support a strong formative use of the item. It was recommended that the item be retained and flagged as difficult, and to add the circle to the description of the level after *n*-gons.
>
> **Item 1021** (*Pi and Circumference* L4: "Knows π is an infinite decimal with approximations of 3.14 or 22/7"); difficulty parameter $b = 2.9$ (predicted: 1.01). It is a T/F item with five statements, which involved seeing if the student thought π was exactly 3.14, 22/7, a nine-digit approximation on a calculator, or knew it lies between two values and "never ends." The data show that many students think π "is the same as" 3.14 (73%) and 22/7 (58%); 88% say π never ends. Only 8% received full credit. The calculator response was rejected by most students. Another item (557) at this level asked students if π is terminating, repeating, and rational; students performed better, with 25% receiving full credit. The learning sciences team, observing that many students gave conflicting answers, such as it is "the same as" to 3.14 and/or 22/7 *and* never ends, wondered if the phrasing of "the same as" was problematic, recommended that the nine-digit calculator display option

be eliminated, and the item be modified to read, for example, "π *is equal to* the decimal 3.14."

Item 526 (*Pi and Circumference* L3: "Names the ratio of circumference to diameter as π"); difficulty parameter $b = 2.1$ (predicted: 0.5). The item shows a "graph" of a set of circles positioned on a y-axis along their diameters, rolling out along the x-axis and mapping to a line plotting circumference vs. diameter (or π). They were asked to identify the ratio represented by the graph. Most (38%) selected diameter:radius. Only 14% recognized it correctly as π. The learning science team questioned the complexity of the graph and of treating the symbol π as an implicit ratio. The decision was to edit the option "π" to "π:1" to use similar construction, and to build another version using a simpler graph and a table.

Item 269 (*Area of Circles* L3: "Use and justify area of circles"); difficulty parameter $b = -0.32$ (predicted: 1.69). It requires a numeric response, asking students to find the area of cell phone receptivity around a radio tower. Fifty-eight percent of the students gave an acceptable response. The most common erroneous response (9%) was 10 (twice the radius). The specific misconception of confusing circumference with area was demonstrated by 5%. Other harder items at the level required students to apply the formula to geometric configurations with partial circles. The learning science team recommended that "and justify" be removed from the level's statement, and that a greater range of items, including, for instance, 2D doughnuts, be included. We decided further to retain this item as a flagged item, which would check for the simplest application of the formula.

Discussion of Q3

Of the five items identified in these two constructs as potentially nonconforming, one (349) was retired as inconsistent with the level. Of the three items flagged for difficulty (453, 1021, 526), one was retained and flagged (453), and two were adjusted (1021, 526). Of the two items flagged for ease (349, 269), one was eliminated and the other was retained and flagged. The adjustments included edits for clarity of the expression, elimination of an option, and minor but important edits to the levels. Discussions included consideration of (1) mathematical specificity and breadth of the levels (Intra-LV), (2) the need to motivate teachers to attend more to certain levels in instruction, and (3) promotion of desirable kinds of mathematical discussions. This analysis provides evidence of the value of conceptualizing the idea of Intra-LV, and the importance of not merely eliminating items whose item difficulty differs from others in the level and may overlap the difficulty of items from upper or lower levels. Further evidence on the performance of these items needs to be collected using think-aloud protocols.

Question 4. After Accounting for Non-conforming Items, to What Extent Do the Construct and the LT Level Explain Item Difficulty, and Can Additional Sources of Variation in Difficulty Be Identified?

Procedure for Q4

After the five items that exhibited non-conformance were eliminated from the data set, the R^2 increased, improving the conformity of the item pool to the expected item difficulty. To help gauge the extent to which the construct and LT theory explain item difficulty, we applied the linear logistic test model (LLTM) (Fischer, 1973) plus error (De Boeck et al., 2011). The model explains Rasch item difficulty in terms of item covariates. The model can be written as

$$\eta_{pi} = \theta_p + \sum_{k=1}^{K} \beta_k X_{(p,i)k} + \varepsilon_i$$

where θ_p is the random person effect, β_k represents the fixed effect of item covariate $X_{(p,i)k}$, and ε_i is the error term. The subscripts p, i, and k represent persons, items, and item covariates, respectively. η_{pi} is a logit link function of the form $ln(\pi_{pi} / (1 - \pi_{pi}))$, where π_{pi} represents the probability of answering an item correctly. Item covariates that produce smaller error variances are considered better predictors of item difficulty than item covariates that produce larger error variances. We first compared the error variances produced from two models: a baseline model that uses only the construct categories to explain item difficulty, and one that includes construct and LT level. To investigate whether we could identify an additional potential source of variance we also compared and a model that includes construct, LT level, and item type as item covariates. We decided to use item type as the third covariate in this model because there is a mixture of item types in our assessment, and research suggests that item types may function differently (e.g., Patz & Junker, 1999). Our primary focus is on the explanatory power of LT level. We used the lmer function of the lme4 package in R to implement this analysis (Bates, Maechler, Bolker, & Walker, 2015).

Results for Q4

The random intercept for items in the first model had a residual standard deviation of 0.79 and a residual variance of 0.62 (Table 4.3). The second model reduced the residual variance to 0.51, representing an 18% reduction in the residual variance. The third model added item type as a covariate and resulted in a residual variance of 0.11. This means that item type accounted for 78% of the unexplained residual variance from the second model. Each fixed effect estimate in Table 4.3 represents the effect that component had on the Rasch item parameter d, which quantifies item "easiness" (it is the multiplicative inverse of item difficulty represented by the b parameter). The fixed effects from the third model show that the covariates with the greatest explanatory influence were the LT levels, which increased the difficulty relative to the other covariates. The levels also generally increased in difficulty in sequential order. Additionally, the *Pi and Circumference* construct decreased the difficulty significantly and Numeric Response items significantly increased the difficulty.

Discussion of Q4

Using the LLTM model to compare models with construct, LT level, and item type as covariates, we found that adding the covariates of LT level and item type to the construct-only model reduced the amount of error variance, suggesting that LT and item type helped explain the variance in item difficulty. In the third model, LT levels were associated with the largest increase in item difficulty and were ordered sequentially in terms of magnitude, which provides further evidence of the relationship between item difficulty and the pattern of levels. The significant decrease in error variance once item type was added suggests that item type and other

Table 4.3 Random Effects: Residual Variances by Model and Fixed Effects of Model 3

Model	Std. Variation	Variance
Model 1: Construct	0.79	0.62
Model 2: Construct and Level	0.71	0.51
Model 3: Construct, Level, and Item Type	0.34	0.11

Fixed Effects of Model 3

	Estimate	Std. Error	z value	Pr(>|z|)
Pi and Circumference	1.23	0.46	2.65	0.008**
Area of Circles	0.15	0.35	0.42	0.68
Level 1	0	0	0	N.A
Level 2	−1.15	0.46	−2.46	0.02*
Level 3	−1.12	0.44	−2.54	0.01*
Level 4	−1.36	0.65	−2.08	0.04*
Level 5	−2.28	0.60	−3.79	0.0001***
Level 6	−2.61	0.56	−4.63	0.000004***
Item type: Numeric Response	−0.56	0.26	−2.13	0.03*
Item type: Multiple Select	−0.97	0.70	−1.37	0.17

*$p < 0.05$; **$p < 0.01$; ***$p < 0.001$

variables, such as context or readability, may be able to give additional insight into what causes variation in item difficulty.

Conclusions

Classroom assessment is emerging as an important addition to the repertoire of assessment practices. Unlike high-stakes assessments, classroom assessments require the return of data in time to inform instructional decisions. Classroom assessments also differ from many benchmark or interim assessments because the feedback from classroom assessments is specific to what is being taught in the curriculum. Compared to many other formative practices, these assessments can contribute to a record of performance over time and across students. Classroom assessments using digital tools can efficiently provide rapid and reliable feedback directly to students and teachers. To do this, it is essential to have evidence that we are measuring what we think we are measuring.

This dynamic context exerts pressure on the validation process. The rapid flow of information and instructional decision making necessitates an iterative process of validation in order to have assessments that can both lead to valuable outcomes for teachers/students (i.e., reliable, "valid" results for the intended use in the instructional flow of a classroom) and inform the creation of future tests (i.e., create more aligned/reliable test forms).

This validation framework reported here is designed to serve the purposes of classroom assessment. It maps the three components (cognition, implementation, and inference) that drove the design and implementation of the system to the six purposes for our classroom assessment. Going forward, this framework provides a means to plan and conduct a set of validation studies aimed at gradually strengthening all aspects of this complex application as it is deployed across many settings.

We reported on the application of our process to a validation argument for a cluster of two learning trajectories. We specified our theoretical view of LTs, and identified three sources of variability and explained how each was to be treated. The approach to Intra-LV, in particular, including the expectation of overlapping ranges of item difficulty for adjacent progress levels, established a principled means by which the step-like structure of the LT is accommodated, cognitively important variation in cases encountered by students is retained, and unnecessary or irrelevant variability is corrected or removed. In response to our research question, we would claim that after modifying the LT based on our protocol, we have increased our confidence that the information provided to teachers concerning the class's progress on the LT is likely to be valid and reliable. We note that the applicability of the study may be constrained because of the low number of items at each level and the small number of field-test sites. Further study of Math-Mapper's use in classrooms is needed.

Several unique factors are brought together in this project. This work represents classroom-level implementation of a coherent system of learning trajectories and associated assessments for an entire grade band, middle school mathematics. The scale of this project (63 learning trajectories in 24 clusters) demands specification of a rigorous iterative validation process. This paper has illustrated our methodology for examining the data one cluster at a time, identifying potential structural anomalies in our measurement approaches, systematically revising and then subjecting them to additional testing. It represents the coupling of a new type of classroom assessment with a process for continually improving the system's key components. Our expectations are, over time, to gradually refine and improve all the LTs, the assessments, and most importantly, the quality of the feedback about students' learning to the key stakeholders—the students and their teachers. By applying a psychometrically valid methodology to the clusters, we believe our approach will systematically provide students and teachers with important information on their progress on LTs, and support growth, as we improve the LTs and their capability to inform instructional decision making.

References

Anderson, J. C., & Gerbing, D. W. (1988). Structural equation modeling in practice: A review and recommended two-step approach. *Psychological Bulletin, 103*(3), 411–423.

Bates, D., Maechler, M., Bolker, B., & Walker, S. (2015). Fitting linear mixed-effects models using lme4. *Journal of Statistical Software, 67*(1), 1–48.

Briggs, D. (2017, November). *Longitudinal growth models and classroom assessment.* Presentation at North Carolina State University, Raleigh, NC.

Cai, L. (2010). High-dimensional exploratory item factor analysis by a Metropolis-Hastings Robbins-Monro algorithm. *Psychometrika, 75*(1), 33–57. https://doi.org/10.1007/s11336-009-9136-x

Carney, M. B., & Smith, E. (in press). Using instrument development processes to iteratively improve construct maps: An example in proportional reasoning. *Journal for Research in Mathematics Education.*

Chalmers, R. P. (2012). MIRT: A multidimensional item response theory package for the R environment. *Journal of Statistical Software, 48*(6), 1–29.

Clements, D. H., & Sarama, J. (2004). Learning trajectories in mathematics education. *Mathematical Thinking and Learning, 6*(2), 81–89. http://doi.org/10.1207/s15327833mtl0602_1

Clements, D. H., Sarama, J. H., & Liu, X. H. (2008). Development of a measure of early mathematics achievement using the Rasch model: The research-based early maths assessment. *Educational Psychology, 28*(4), 457–482. http://doi.org/10.1080/01443410707777272

Common Core State Standards Initiative. (2010). *Common Core State Standards for mathematics.* Washington, DC: National Governors Association Center for Best Practices, Council of Chief State School Officers. Retrieved from www.corestandards.org/wp-content/uploads/Math_Standards.pdf

Confrey, J. (2015). Some possible implications of data-intensive research in education: The value of learning maps and evidence-centered design of assessment to educational data mining. In C. Dede (Ed.), *Data-intensive research in education: Current work and next steps* (pp. 79–87). Washington, DC: Computing Research Association.

Confrey, J., Maloney, A. P., & Nguyen, K. H. (2014). Learning trajectories in mathematics: Introduction. In A. P. Maloney, J. Confrey, & K. H. Nguyen (Eds.), *Learning over time: Learning trajectories in mathematics education* (pp. xi–xxii). Charlotte, NC: Information Age Publishing.

Confrey, J., Maloney, A. P., Nguyen, K. H., & Rupp, A. A. (2014). Equipartitioning, a foundation for rational number reasoning: Elucidation of a learning trajectory. In A. P. Maloney, Confrey, J., & Nguyen, K. H. (Eds.), *Learning over time: Learning trajectories in mathematics education* (pp. 61–96). Charlotte, NC: Information Age Publishing.

Confrey, J., McGowan, W., Shah, M., Belcher, M., Hennessey, M., & Maloney, A. (2019). Using digital diagnostic classroom assessments based on learning trajectories to drive instruction and deepen teacher knowledge. In D. Siemon, T. Barkatsas, & R. Seah (Eds.), *Researching and using learning progressions (trajectories) in mathematics education*. Rotterdam: Sense Publishers.

Confrey, J., Toutkoushian, E., & Shah, M. (2019). A validation argument from soup to nuts: Assessing progress on learning trajectories for middle school mathematics. *Applied Measurement in Education*.

de Ayala, R. J. (2009). *The theory and practice of item response theory*. New York, NY: The Guilford Press.

De Boeck, P., Bakker, M., Zwitser, R., Nivard, M., Hofman, A., Tuerlinckx, F., & Partchev, I. (2011). The estimation of item response models with the lmer function from the lme4 package in R. *Journal of Statistical Software*, *39*(12), 1–28.

Fischer, G. H. (1973). Linear logistic test model as an instrument in educational research. *Acta Psychologica*, *37*(6), 359–374. https://doi.org/10.1016/0001-6918(73)90003-6

Graf, E. A., & van Rijn, P. W. (2016). Learning progressions as a guide for design: Recommendations based on observations from a mathematics assessment. In S. Lane, M. R. Raymond, & T. M. Haladyna (Eds.), *Handbook of test development* (2nd ed., pp. 165–189). New York, NY: Taylor & Francis.

Haertel, E. H., & Lorie, W. A. (2004). Validating standards-based test score interpretations. *Measurement*, *2*(2), 61–103. https://doi.org/10.1207/s15366359mea0202_1

Hatcher, L. (1994). *A step-by-step approach to using SAS for factor analysis and structural equation modeling*. Cary, NC: SAS Institute.

Kane, M. T. (2006). Content-related validity evidence in test development. In M. T. Haladyna & M. S. Downing (Eds.), *Handbook of test development* (pp. 131–153). Mahwah, NJ: Lawrence Erlbaum Associates, Inc.

Kolen, M. J., & Brennan, R. L. (2004). *Test equating, scaling, and linking*. New York, NY: Springer.

Lai, E. R., Kobrin, J. L., DiCerbo, K. E., & Holland, L. R. (2017). Tracing the assessment triangle with learning progression-aligned assessments in mathematics. *Measurement: Interdisciplinary Research and Perspectives*, *15*(3–4), 143–162. https://doi.org/10.1080/15366367.2017.1388113

Lehrer, R., Kim, M. J., Ayers, E., & Wilson, M. (2014). Toward establishing a learning progression to support the development of statistical reasoning. In A.

P. Maloney, J. Confrey, & K. H. Nguyen (Eds.), *Learning over time: Learning trajectories in mathematics education* (pp. 31–59). Charlotte, NC: Information Age Publishing.

Lehrer, R., & Schauble, L. (2015). Learning progressions: The whole world is NOT a stage. *Science Education, 99*(3), 432–437. https://doi.org/10.1002/sce.21168

Mason, A. J. (2012). OpenSolver: An open source add-in to solve linear and integer programmes in Excel. In D. Klatte, H.-J. Lüthi, & K. Schmedders (Eds.), *Operations research proceedings 2011* (pp. 401–406). Berlin, Germany: Springer. http://dx.doi.org/10.1007/978-3-642-29210-1_64

Messick, S. (1989). Validity. In R. L. Linn (Ed.), *Educational measurement* (pp. 13–103). New York, NY: MacMillan.

Mislevy, R. J., Steinberg, L. S., & Almond, R. G. (2003). On the structure of educational assessments. *Measurement: Interdisciplinary Research and Perspectives, 1*(1), 3–67. https://doi.org/10.1207/S15366359mea0101_02

National Council on Measurement in Education. (2017). *Special conference on classroom assessment and large-scale psychometrics: The twain shall meet.* Lawrence, KS.

National Research Council. (2001). *Knowing what students know: The science and design of educational assessment* (J. W. Pellegrino, N. Chudowsky, & R. Glaser, Eds.). Washington, DC: National Academies Press.

Nichols, P. D., Kobrin, J. L., Lai, E., & Koepfler, J. (2016). The role of theories of learning and cognition in assessment design and development. In A. A. Rupp & J. P. Leighton (Eds.), *The handbook of cognition and assessment: Frameworks, methodologies, and applications* (pp. 15–40). West Sussex, UK: John Wiley & Sons.

Patz, R. J., & Junker, B. W. (1999). Applications and extensions of MCMC in IRT: Multiple item types, missing data, and rated responses. *Journal of Educational and Behavioral Statistics, 24*(4), 342–366. https://doi.org/10.3102/10769986024004342

Pellegrino, J. W., DiBello, L. V., & Goldman, S. R. (2016). A framework for conceptualizing and evaluating the validity of instructionally relevant assessments. *Educational Psychologist, 51*(1), 59–81. https://doi.org/10.1080/00461520.2016.114550

Pham, D., Bauer, M., Wylie, C., & Wells, C. (in press). *Using cognitive diagnosis models to evaluate a learning progression theory.*

Shepard, L. A., Penuel, W. R., & Pellegrino, J. W. (2018). Using learning and motivation theories to coherently link formative assessment, grading practices, and large-scale assessment. *Educational Measurement: Issues and Practice, 37*(1), 21–34. https://doi.org/10.1111/emip.12189

Wilmot, D., Schoenfeld, A., Wilson, M., Champney, D., & Zahner, W. (2011). Validating a learning progression in mathematical functions for college readiness. *Mathematical Thinking and Learning, 13*(4), 259–291. https://doi.org/10.1080/10986065.2011.608344

Wilson, M. (2005). *Constructing measures: An item response modeling approach.* Mahwah, NJ: Lawrence Erlbaum Associates, Inc.

5 Assessing College-Ready Data-Based Reasoning

Amy Arneson, Diah Wihardini, and Mark Wilson

We have seen in recent years growing emphasis on preparing college-ready students in K-12 education. A very significant example is that the Common Core State Standards (National Governors Association Center for Best Practices [NGACBP], & Council of Chief State School Officers [CCSSO], 2010) explicitly names college readiness as one of its goals. To be college-ready, a high school graduate generally needs to be equipped with the prerequisite skills and knowledge to be able to understand and cope with the content knowledge presented in a course taken during the initial part of a college degree, and have the dispositions to succeed in postsecondary academic and social life (Conley, 2007). These circumstances have motivated a call for increased rigor and explicit focus on college readiness (CR) during K-12 schooling and in turn the development of valid, reliable, and psychometrically sound measures of CR. Though the ACT and SAT have long been considered correlational indices of CR, the content of these entrance exams is very limited and they cannot be considered complete measures of CR. "College readiness" is most often defined multidimensionally, cross-cutting all academic domains and including many "non-cognitive" skills (Conley, 2007; Wiley, Wyatt, & Camara, 2011). It is important when developing measures of CR to be explicit about what aspects of CR will be measured.

In this chapter, we focus on our work on *college readiness in data-based reasoning*.[1] This is the critical statistical and data-based reasoning required for first-year university coursework *outside* of mathematics and statistics departments. Thus, we have developed assessment tasks that are representative of the types of things expected of students in introductory courses in natural sciences, social sciences, humanities, arts, health care, and others who are not mathematics or statistics majors. Thus, this encompasses the proportion of college students who use mathematics in the course of their studies rather than those who go on to major in mathematics or statistics. This chapter describes the design process and data analysis for the Critical Reasoning for College Readiness (CR4CR) Assessment. This work is not focused on the mathematical mechanics of statistics but its application in entry-level coursework outside of mathematics and statistics departments.

The methods we use for test development follow the BEAR Assessment System (BAS) (Wilson & Sloane, 2000; Wilson, 2005), a principled approach to the development and validation of measurements. First, we give a broad working definition of college-ready data-based reasoning and provide details on two specific elements, called constructs, that we aim to measure and that are used as examples in this chapter. The next section provides some example assessment tasks aligned to these constructs. BAS, its relationship to validation, and the statistical modeling methods are presented in the third section. The fourth section gives data analysis results from a field test of the CR4CR Assessment, and the fifth and final section discusses future directions of this research.

A Framework for College-Ready Data-Based Reasoning

Some level of critical reasoning skill is needed for everyday life, if just to understand some of the headlines that pop up on social media. Information from sources of varying quality about election-year polling results, drug trials, the cost of college, life expectancy, and so forth is readily available and commonplace. Graphics and charts that are circulated on the internet may be deliberately designed to mislead. Even reputable news outlets have been known to circulate misinterpretations of statistical findings when the writer or editor is misinformed or underinformed (von Roten, 2006; Newman, 2012).

Watson and Callingham (2003) present a definition of "literacy" that includes numeracy (aka quantitative literacy) as well as data handling and chance. The quantitative aspects of literacy are growing increasingly important with the rise of data science (specifically "big data"), the availability of powerful analysis software, and the use of statistical findings in everyday media. College-bound students are hard-pressed to choose a major that will not require them to draw conclusions from data, critique an article that includes some presentation of data, or use data analysis results as evidence for an argument (Conley, 2005).

In the Standards for Success (Conley, 2005) and Reaching the Goal (Conley, Drummond, de Gonzalez, Rooseboom, & Stout, 2011) research, college professors, especially those in the natural and social sciences, indicated that prerequisite knowledge of basic statistical concepts and techniques plays an important role in entry-level courses. Conley (2005) reports that students who fail initial coursework because they lack a prerequisite skill will avoid majors in that area of study, "closing off entire avenues of the curriculum and career pathways" (p. 114). An interesting finding of this research is that statistical skills are less important in entry-level coursework for mathematics majors, while they are imperative for success in other majors in the natural sciences (e.g., biology, ecology, physics), the social sciences (e.g., economics, psychology, journalism), professional degree programs (e.g., nursing), and the humanities (e.g., history).

A survey of Advanced Placement (AP) course and exam descriptions that the authors carried out in a previous study indicated that some level of statistical knowledge was a prerequisite for most of the AP courses offered (Wilson, Arneson, Mason, & Wihardini, 2017). AP coursework was surveyed in this study as it can be considered reflective of entry-level coursework[2] and provide an easily accessible source of widely used documentation. Further, it provided us with specific example tasks to supplement the conclusions from the Conley (2005) and Conley et al. (2011) research, which were drawn regarding broad topic areas and standards, to guide our construct and item development. Thirty-three[3] AP courses were surveyed, and statistical or data analysis skills were either referenced in the course objectives or included in sample items in 28 of the 33 surveyed courses. In both the Conley research and our AP survey, some of the commonly encountered skills related to statistics, probability, and data were:

- Reading, interpreting, and creating graphical displays of data;
- Collecting and organizing data from experiments and sample surveys;
- Summarizing data numerically;
- Comparing distributions of data;
- Drawing formal inferences from data;
- Identifying trends in data;
- Critically evaluating reports on data.

We also looked at other recent attempts made to identify the skills and knowledge required for students to be considered college-ready. Some of the most notable collections of CR standards are the Common Core State Standards–Mathematics (CCSS-M) (NGACBPC & CCSSO, 2010); Knowledge and Skills for University Success (KSUS) (Conley, 2005); and College Board Standards for College Success–Mathematics and Statistics, Adapted for Integrated Curricula (Mathematics and Statistics Standards Advisory Committee, 2007). Along with these three documents, three other collections of K-12 academic standards (not necessarily focused on CR) for mathematics and statistics were surveyed: Mathematics Framework for the National Assessment of Educational Practices (National Assessment Governing Board, 2013); Principles and Standards for School Mathematics (NCTM, 2000); and Guidelines for Assessment and Instruction in Statistics Education (GAISE) Report: A Pre-K-12 Curriculum Framework (Franklin et al., 2007).

Combining the information garnered from the sources described above, we have developed a *Framework for College-Ready Data-Based Reasoning*, which consists of three high-level essential questions that a student should ask and answer, regardless of the course content or research context:

1. What decisions need to be (were) made based on the data? (Decision Making)

2.   Where did the data come from? (Producing and Selecting Data)
3.   What was done to the data? (Summarizing Data)

This is illustrated in Figure 5.1, each question constituting a vertex of a triangle and with probability in the center, as probability theory underlies the concepts contained in the other three areas. For the CR4CR Assessment, however, we have not yet formulated constructs targeted at probability theory because we are primarily concerned with applied contexts. The traditional mathematical study of probability often removes the situational context. We plan to introduce a construct aimed at contextualized and practical probability knowledge in a second round of effort.

We chose to use the existing standards and frameworks to inform our own, as none on their own thoroughly defined data-based reasoning. Each, though, illuminated elements of data-based reasoning that we incorporated into our framework. Specifically, we considered the contexts of (1) reading and critiquing reports on data and (2) planning a data-based investigation, two common tasks for which we posit parallel thinking processes. For the purposes of this chapter, we frame the discussion in the first context of critiquing reports on data, as most of the CR4CR Assessment items are written in this context. The work of Conley et al. (2011) reports that the standard "Evaluate reports on data" of the CCSS-M was rated as one of the most important standards in the statistics strand by instructors of first-year courses. Lajoie and Romberg (1998) state that K-12 students should learn to both critique as well as produce reports of statistical results "as required" for their future roles as consumers and produces in society.

Framework for College-Ready Statistical Thinking

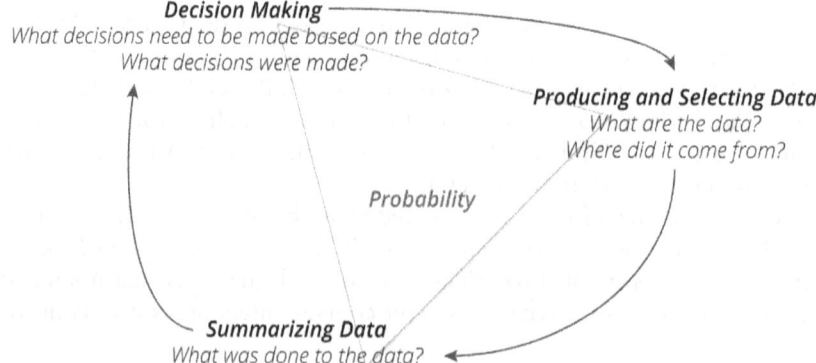

Figure 5.1 A framework for college-ready data-based reasoning

Although it makes sense to see each of the three broad questions as an individual part in a sequence of steps in an investigation—formulate a question, get the data, analyze the data—the three parts need to also be considered simultaneously. For example, it makes no sense to think about the qualities of data collection without also considering the use of those data for decision making. A data-based investigation starts with the motivation—a question is formulated by first answering *what decisions need to be made based on data?* From there, it moves to the selection and production of data—*What are the data and where did they come from?* Once data have been obtained, they are analyzed—*What was done to the data?* That analysis then informs the claim or decision made, bringing the investigation back to the first critical question. Often, the investigation stops here. However, some may iterate through this critical questioning process by either refining steps of the investigation or repeating them. We conceptualize data-based reasoning as the practice of asking oneself these essential questions when presented with a report on data. Note, however, that these are very broadly defined, and the critical thinker does not stop with these. A cascade of questions will arise based on these broad, starting questions. What the follow-up questions will be is largely determined by the material presented to the student and the domain in which the context exists.

This categorization of skills is not new. Schaeffer, Watkins, and Landwehr (1998) organize the statistical content they felt should be in the K-12 curriculum into these same categories with the addition of "number sense." Their "Planning a Study and Producing Data" strand is our "Producing and Selecting Data," "Data Analysis" is "Summarizing Data," and "Inferential Reasoning" is "Decision Making." They claim that the strands can be roughly ordered by level of sophistication and intellectual difficulty, which we do not. In our framework, constructs are situated within these different categories, but we have not yet theorized how each construct is related to the others. Further, we contend that all of these broad critical questions that define our categories should be considered simultaneously, and thus there is probably great overlap in terms of difficulty between constructs in the different categories. Though we do not propose a novel structure to K-12 statistics, our framework is focused on statistical and data-based reasoning, not the whole of the study of statistics.

As such, the conception of critical questioning for statistical literacy (Watson & Callingham, 2003) underlies the selection and illustration of constructs we chose to include in this framework. Watson's work is grounded in a three-tier framework for statistical literacy (Watson, 2006) in which each tier describes an increased sophistication in thinking about a statistical problem. This complements our construct modeling approach which also defines hierarchies of development in critical statistical literacy skills, though we aim to flesh out more specific elements of statistical literacy and with a refined focus on college readiness, not general citizen statistical literacy.

In addition to the mathematics and statistics standards already referenced, we also looked for elicitation of data-based reasoning in the College Board Standards for College Success for Science (Science Standards Advisory Committee, 2009), the Next Generation Science Standards (NGSS) (NGSS Lead States, 2013), and the KSUS Standards in all other disciplines (English, Natural Sciences, Social Sciences, Second Languages, and Arts). This was an especially important task as our main area of inquiry is in how statistics is applied outside of the mathematics-based disciplines. These standards, along with the AP course survey and Conley et al. (2011) research findings, guided the decisions about which of the subtopics/skills should be the focus in the initial development of the CR4CR Assessment. The resulting list of skills was then further refined into single, potentially measurable variables (aka constructs). In this chapter, we present two of these constructs defined in full. We do not claim to be covering the entire breadth of data-based and statistical reasoning with these constructs. These will serve as demonstration constructs and are the beginning to laying out a more complete conception of college-ready critical reasoning.

Two Data-Based Reasoning Constructs

The two demonstration constructs described in this chapter, Linking Data to a Claim (LDC) and Formal Inference (FoI), are both aligned to more specific critical questions situated under an essential question of our framework. To contextualize this abstract discussion about psychometric constructs, consider an example. Suppose a magazine article contains the statement "Women feel more positively about the idea of paid paternity leave in the workplace than men." The phenomenon under investigation, here, is the relationship between gender and opinion toward paternity leave. Suppose the claim is based on the following data: responses to two multiple-choice questions ("What is your gender?" and "Do you think employers should offer paternity leave to new fathers?") from a voluntary e-mail survey sent to the magazine's subscribers. A t-test for the difference in proportions of "Yes" answers to the paternity leave survey question between men and women was conducted and the difference was statistically significant at the usual ($\alpha = 0.05$) level. There are many critical questions that a statistical thinker should be asking herself, such as the following:

- Are these data appropriate for the phenomenon under study?
- Is a t-test for the difference in proportions the appropriate statistical test?
- Is the result of practical importance?

• Does the claim overreach in terms of the population to which it generalizes?

The first question exemplifies the LDC construct and the second and third critical questions are aligned to FoI. The fourth is related to a construct in our framework but not discussed in this chapter, called Evaluating the Limitations of a Claim (ELC, available from the authors by request). We chose the LDC and FoI constructs for demonstration in this chapter because they represent both novel and familiar conceptions in the literature surrounding defining and measuring statistical reasoning constructs. LDC is at its core an argumentation construct, a topic that is receiving more attention in the science education literature along with the release of the NGSS, and we believe that it should also be considered in the context of data science and statistics education. FoI is related to a more familiar statistical reasoning skill, interpretation and understanding of inference statistics. Our construct map draws upon inspiration from the Informal Inference construct developed for the Modeling Data curriculum (Lehrer, Jones, Kim, Pfaff, & Shinohara, 2014).

To show how these questions are operationalized in the CR4CR Assessment, Figure 5.2 provides a sample task called *Miles per Gallon* (MPG). The test was delivered on a computer, using the BEAR Assessment System Software (Torres Irribarra, Freund, Fisher, & Wilson, 2015). Figure 5.2 shows the rendering of the task as it would be on the screen of a student taking the test, with some sample student responses provided in the text input boxes in italics. Items 1–3 of MPG are aligned to FoI, while item 4 is an LDC item. The item design process is described in more detail in a later section.

Linking Data to a Claim (LDC)

In our examination of standards and courses, we found that much of the data-based reasoning involved in first-year coursework tasks is not explicitly quantitative. Though quantitative statistical procedures are routinely conducted to test hypotheses, we should not lose sight of the real-world phenomena that any given statistical study is meant to represent. An important task in either planning a statistical study or critically evaluating one is to justify the link between the real-world object under study and the variables or measurements that are purported to reflect what we want to know about that object—these measured variables make up the data and a student must ask herself: *Are these data the right data?* Explicit consideration of this connection may not be a concern in traditional statistics courses when students learn statistical procedures and the data are supplied by the textbook. In these settings, it is often taken for granted that the data collected are the right data to test the hypothesis. However, in applied settings, this concern, that the

An automotive industry lobbyist wishes to claim that cars have been getting more energy efficient over time, and thus the federal government should relax some of the environmental regulations they impose in the industry.

To back up this claim, they use Environmental Protection Agency (EPA) fuel economy data to compare the "city" miles per gallon (MPG) ratings of random samples of cars from 2012, 2013, and 2014. Using statistical software, they found the mean MPG along with 95% confidence intervals. These are reported in the following table.

	Mean MPG	95% confidence interval
2012	19.41	(18.01, 20.81)
2013	20.73	(19.40, 22.06)
2014	22.03	(20.84, 23.22)

[1] The lobbyist makes the claim "There was a statistically significant improvement in fuel efficiency **from 2012 to 2013**." Do you agree with this statement? Why or why not?

No, because the mean MPG in 2012 could be as high as 20.81 and the mean in 2013 could be as low as 19.40. If that's the case, the fuel efficiency actually got worse.

[2] The lobbyist also makes the claim "There was a statistically significant improvement in fuel efficiency **from 2012 to 2014**." Do you agree with this statement? Why or why not?

Yes, the mean is 2.62 higher.

[3] Is the observed 2.62 MPG improvement **from 2012 to 2014** in fuel economy large enough to be of importance to an average driver? Why or why not?

I would say not as the difference is only 2.62 miles per gallon. This means that you only save about 3 cents per mile so you would have to drive a long way to really notice.

[4] Recall that the lobbyist wants to use this analysis as evidence for a claim that government regulations can be relaxed. *Without regard to the analysis results*, do you think that he chose the right data to analyze? Why or why not?

No, environmental impact is more complicated than that.

Figure 5.2 The *Miles per Gallon* (MPG) item cluster

selected data truly reflect the phenomena about which a claim is made, is of central importance.

The LDC construct map describes development in reasoning about the link between the data and the content of a statistical claim (and/ or the research question that triggers an investigation). We acknowledge that many frameworks for statistical investigation and statistical learning include a step for formulating a research question (e.g., Schaeffer et al.,

1998), so it is important to note that the levels of this construct are conditioned on that question already having been formulated.

Critical evaluation of a claim is an exercise in argumentation; students must evaluate the evidence behind that claim. Suppose the survey in the paternity leave context described above also contained the question, "Did you take maternity/paternity leave after the birth of your last child?" If the responses to that question were chosen as the data to analyze, there would be discontinuity between the observations and the content of the claim and thus would not be the right data for this context. Student recognition that some data are better than others is essential in applied settings, and it is crucial that they can critique the choices of others or present an argument for their choices.

The LDC construct map is shown in Figure 5.3. The qualitative descriptions of each level are provided in the left side of the table with sample responses to question 4 of the MPG task (provided in Figure 5.2) in the rightmost column. It should be read from the bottom up, as that is the direction of increasing sophistication within the construct. Each successive level represents a qualitatively different type of student response to an item, indicating an increased level of sophistication in thinking. We observe that students tend to develop gradually from lower levels to higher levels, and we use these ideal points as milestones along this continuum of development: not all students will fit distinctly onto one of these points; the majority will be somewhere in between them.

This construct is an integration of the argumentation construct map developed by Henderson, Osborne, MacPherson, and Szu (2014) and the structure of observed learning outcomes (SOLO) taxonomy (Biggs & Collis, 1982). The SOLO taxonomy outlines understanding of a generic concept with increasing levels of coordination, from the ability to simply identify elements to the ability to integrate those elements. For LDC, we used SOLO to expand the Henderson et al. (2014) map. Their model for argumentation is structured around the three elements of an argument— claim, evidence, and warrant—and the coordination of these three elements and then of evaluating competing arguments. In our context, a statistical claim is made, and the data collected are the evidence. The warrant then, is the argument that the data are appropriate data for the claim. For the LDC construct, we are focused only on the warrant element, and we refer to this as the link between the data and the claim. Similar to Henderson et al. (2014), we compare unilateral (one-sided) with balanced (two-sided) justifications for data/observation selection (i.e., consideration of both arguments *for* and *against*).

The bottom of the construct, LDC0, characterizes a respondent that does not identify a statistical claim. Moving up, at the Unevidenced level (LDC1), the respondent identifies or provides a claim, but without accompanying evidence. Students at this level fail to recognize there is a question about the choice of data at all. At the External level (LDC2),

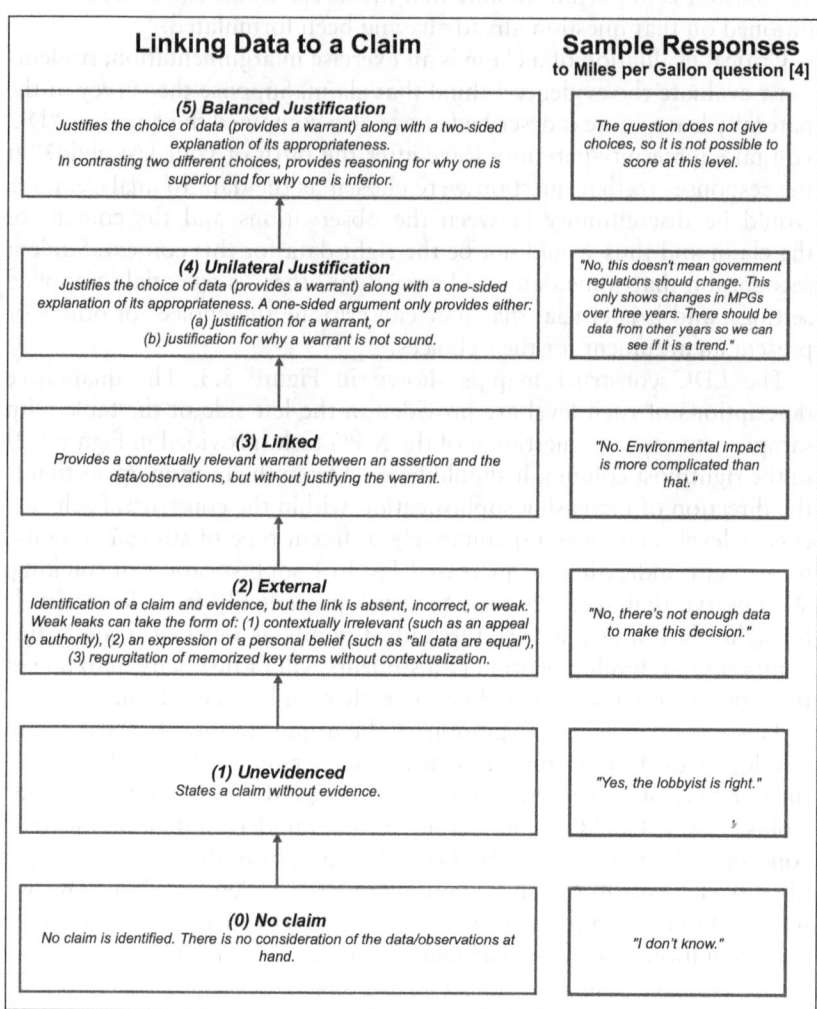

Figure 5.3 The Linking Data to a Claim (LDC) construct map

respondents may judge the appropriateness of the data collected by rely-
ing only on an external (contextually irrelevant) authority or their own
beliefs about the subject at hand, or they may hold an "always or never"
attitude—those who seem to think that the inclusion of a number, any
number, in a claim is evidence enough for that claim, or those who trust
no number.

LDC3 (Linked) is the first in which the link between the data and
the claim is explicitly drawn by the respondent. However, the link is
merely identified and there is no justification or statement for it as a

choice. Respondents may not recognize that there is a choice of observations to represent some phenomena of interest. This recognition may be taken for granted, especially for tasks that involve physical/manifest phenomena that are easily measured. Consider question 4 of the MPG task (Figure 5.2). It may be taken for granted that average miles per gallon represents a car's impact on the environment, but we expect higher-level students to think about the bigger picture and recognize that other things can affect a car's environmental impact.

Respondents at LDC4 start to realize that sometimes a link is justified and sometimes it is not. This is where the credibility of the data becomes a prime issue—the warrant is not only predicated on the existence of a link, but also that the data themselves are credible in this context. Of course, some data, such as randomly generated numbers, may be credible in only unusual cases, so that they may be unsuitable for a whole range of claims. However, they provide single-sided justifications only, focusing only on the choice that was made. Responses at the highest level, LDC5, provide more complex and balanced justifications of the choice of data to represent the phenomena of interest. They compare choices and provide reasoning in favor of the choice and against other choices. The distinction between LDC4 and LDC5 is the inclusion of critique. LDC4 is simply an argument for (or against) a choice of data. LDC5 furthers the argument by critiquing an alternate choice(s) by contrasting it with the respondent's choice of data. An LDC4 response is an argument that may be of the structure "It is good because ___." At LDC5, the argument takes a comparative form: "It is better than the alternative(s) because ___."

Formal Inference (FoI)

An important part of interpreting statistical results is determining their significance (aka conducting inference). Formal procedures that produce confidence intervals, p-values, or effect sizes are important elements of many first-year courses in the sciences, as many fields of research rely heavily on them. Students need to be critical of the use of the term "significant" in articles and reports, because it could be used to refer to statistical significance, practical significance, or the common, non-technical definition of significance, which refers to its relevance to the topic. Statements of statistical significance usually accompany the results of formal statistical tests (e.g., t-tests, confidence intervals); practical significance is often quantified by an effect size (Agresti & Finlay, 2009).

Understanding of the logic of inference and how it leads to conclusions of significance has traditionally been downplayed in statistics education, but there has recently been more attention paid to this issue with the increasing availability of computer-based simulation and other technology-enhanced learning tools for the classroom (GAISE College Report ASA Revision Committee, 2016). For students to understand what they are doing when they conduct their own statistical tests, and to understand the results presented in media reports or scholarly works, more focus may need to be given to this area of statistical reasoning. This is especially important in the natural and social sciences. Recent work on science standards have recognized this explicitly. One of the NGSS goals is that "By grade 12, students should be able to analyze data systematically, either to look for salient patterns or to test whether data are consistent with an initial hypothesis" (NGSS Lead States, 2013).

The logic of significance is a particularly difficult thing for students to understand, as it requires reasoning about conditional statements, for which their understanding is often flawed (Evans, Newstead, & Byrne, 1993). Typically, the reasoning of statistical significance tests starts with the assumption that some effect of interest is not present in the population (the condition). Evidence for the effect (or evidence against the absence of an effect) is established by a low probability of obtaining the observed result under the assumption that there is no effect. This type of reasoning using a conditional statement in which negation of the conclusion leads to negation of the condition is called modus tollens, which has been found to be a difficult type of reasoning for people to master (Evans et al., 1993). The logic of inference from confidence intervals is similar (Henriques, 2016). Even though the logic/reasoning is difficult, we found that first-year college students are exposed to the results of statistical tests in applied coursework.[4]

While the interpretation of statistical significance is generally uncontroversial in the statistics literature, guidelines for defining, using, and interpreting practical significance in decision making processes is less well agreed upon (Kelley & Preacher, 2012). It is widely recommended by research organizations that effect sizes be reported alongside determinations of statistical significance (e.g. American Psychological Association, 2010; Task Force on Reporting of Research Methods in AERA Publications, 2006; National Center for Education Statistics, 2002) and we take the stance that university coursework should be held to the same standard. An understanding of practical significance is important for reporting and critically reading statistical results, and the higher levels of our proposed construct include the formalization of this understanding in terms of an effect size. We follow the definition for effect size given by Kelley and Preacher (2012): "a quantitative reflection of the magnitude of some phenomenon that is used for the purpose of addressing a question of interest" (p. 140). This definition is suitable for our construct as it describes development through the thinking process of considering the context of the statistical investigation (i.e., the question being addressed) in the choice of effect size that is reported. The highest level of our construct is the integration of practical and statistical significance, where a critical thinker balances the information provided by each and integrates *both* sources of information about the research question at hand, especially when they seemingly provide conflicting information (e.g., if a result is statistically but not practically significant or vice versa). We do not expect an incoming university freshman student to have a complete library of types of effect sizes, but rather to understand the logic behind them and, as Kelley and Preacher describe, be able to choose "relevant effect sizes" with subjective input from field experts. We are looking to assess whether and how students are thinking about these choices, and hope to emphasize the importance of training statistical thinkers to not make ad hoc choices about these things and to think critically about the choices that are made by others.

We propose a double branched construct map for FoI, shown in Figure 5.4. This structure differentiates the progression of student development of claims concerning significance related to (a) statistical significance (frequentist statistical tests and confidence intervals) and (b) practical significance (effect size). The lower levels in both branches are identical. The split occurs after FoI2. It is important to note, then, that after the split we do not necessarily expect students to progress on the two branches at equal rates. At this point, there is no claim being made about the ordering of the levels across the two branches. However, we do contend that the integration of the two branches constitutes the highest level, Integrated Inference (FoI5). We do not focus on the procedures of conducting statistical tests or calculating effect sizes but, rather, the use and interpretations of the results that they produce. This construct instead maps the development of informed decision making using the results of inference procedures. In a way, it asks, "To what extent is a conclusion (or claim) informed by the statistical procedure?"

The FoI construct map is intended to describe the developmental progression through the two branches of formal inference. It may not be necessary for students to reach the topmost level of this construct to be prepared for first-year coursework. Correct interpretation of statistical tests, confidence intervals, and effect sizes may be sufficient for the critical evaluation of reports on data analysis that is expected in first-year coursework. However, out of concern for completeness, we have drafted the construct to the highest theorized level.

FoI0, no inference, characterizes a response that does not contain a claim. Responses at the FoI1 level where an inference is made do not consider the idea of significance—the result of the data analysis is directly used to make the claim. For example, if the treatment group in a randomized drug trial has a lower mean blood pressure than the control group, this difference is reported and interpreted as such. The localized result is all that is reported, and it may be reported at the population level. At FoI2, respondents rely on their own personal intuitions about the magnitude of the results to make a determination of significance. For instance, an FoI2 response may be along the lines of "One inch isn't that big, so the difference is not significant." This response fails to contextualize their conclusion about how significant

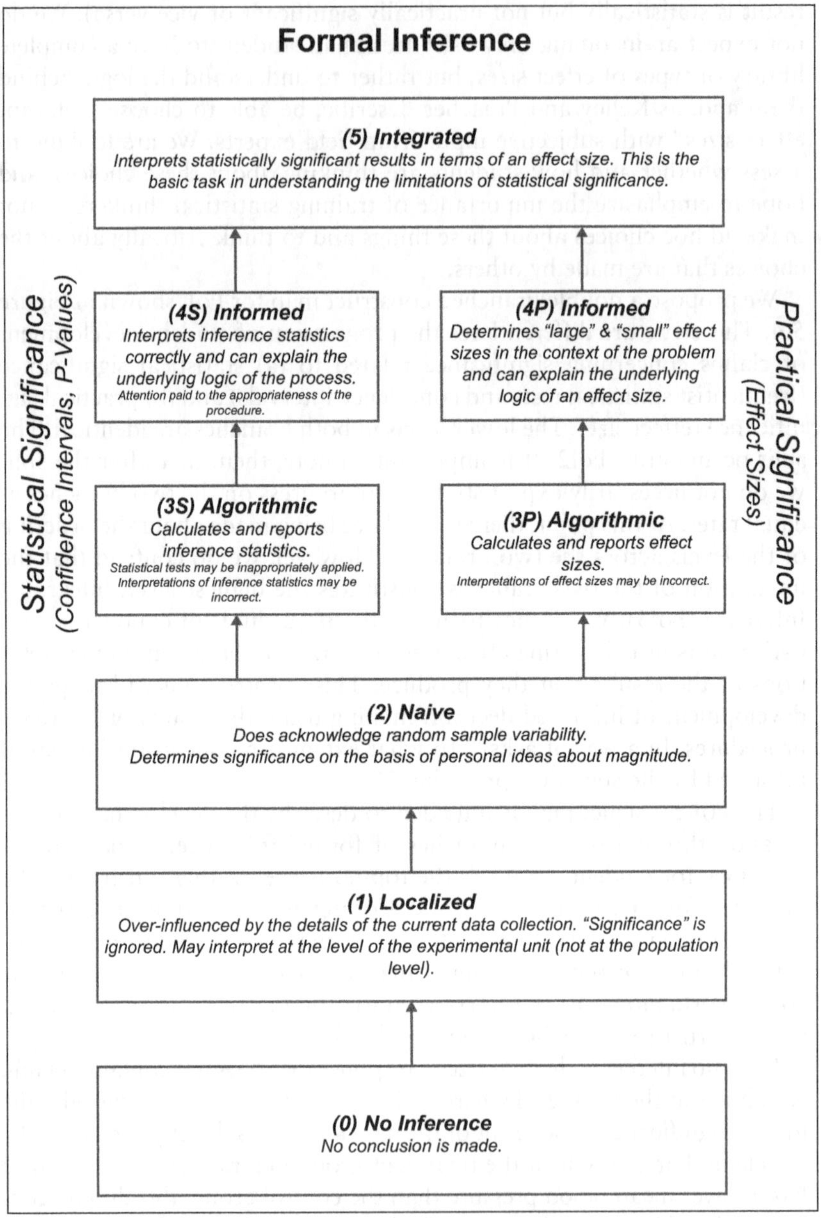

Formal Inference

(5) Integrated
Interprets statistically significant results in terms of an effect size. This is the basic task in understanding the limitations of statistical significance.

(4S) Informed
Interprets inference statistics correctly and can explain the underlying logic of the process.
Attention paid to the appropriateness of the procedure.

(4P) Informed
Determines "large" & "small" effect sizes in the context of the problem and can explain the underlying logic of an effect size.

(3S) Algorithmic
Calculates and reports inference statistics.
Statistical tests may be inappropriately applied. Interpretations of inference statistics may be incorrect.

(3P) Algorithmic
Calculates and reports effect sizes.
Interpretations of effect sizes may be incorrect.

(2) Naive
Does acknowledge random sample variability. Determines significance on the basis of personal ideas about magnitude.

(1) Localized
Over-influenced by the details of the current data collection. "Significance" is ignored. May interpret at the level of the experimental unit (not at the population level).

(0) No Inference
No conclusion is made.

Statistical Significance
(Confidence Intervals, P-Values)

Practical Significance
(Effect Sizes)

Figure 5.4 The Formal Inference (FoI) construct map

one inch is, relative to the phenomenon under study. For example, one inch of growth in the height of a toddler is relatively more meaningful than a one-inch difference in the distance of a daily commute to work, hence context matters for practical significance. Respondents who use the term significance colloquially, and don't differentiate between statistical and practical significance, would be placed here. Gross misinterpretations of inference statistics would also be placed here.

We will discuss the two branches of the construct together, as they parallel each other in explanation. At the Algorithmic levels (FoI3P and FoI3S), a respondent can execute the algorithm to produce an inference statistic. *P*-values or standardized effect sizes are reported, but there may be subtle misinterpretations of these values, such as interpreting a *p*-value of 0.02 as "The probability of my data is 2%" or interpreting a 95% confidence interval as an interval that has a 95% probability of containing the true value. It is important to note, again, that we do not necessarily expect respondents to be at the same levels for the two branches. They may very well have the algorithm to produce a 90% confidence interval, placing them at FoI3S, but not know a thing about Cohen's *d*, not quite reaching FoI3P. At these levels, procedures may be inappropriately applied and inference may be drawn about incorrect populations, typically as overgeneralization.

However, at FoI4P and FoI4S, attention is paid to the procedure because this is when respondents can explicate the logic behind statistical tests. In order to do so in context, the data generating model of the null hypothesis is understood and thus the correct procedure to match the data structure and research question must be chosen. We call this *informed* determination of statistical (FoI4S) or practical (FoI4P) significance. Finally, at FoI5, respondents coordinate both statistical and practical significance. Here, the limitations of *p*-values are recognized in the context of effect sizes. Both are reported with correct interpretations and they are coordinated.

The BEAR Assessment System (BAS)

The BEAR Assessment System (Wilson & Sloane, 2000) is an assessment development approach grounded in four principles: (1) a developmental perspective on student learning, (2) a match between instruction and assessment, (3) management by teachers, and (4) generation of high-quality reliability and validity evidence. It can be used as a method to develop learning progressions for assessment purposes (Wilson, 2009) and thus is the approach we took for the CR4CR Assessment. The four principles are embodied in the four building blocks, described below, which are the principled steps that we took in order to construct the CR4CR Assessment. These building blocks are iterative and integrated, meaning that the cycle is usually completed multiple times and the building blocks each inform one another. Following BAS is helpful in building a validation argument for an instrument as the steps directly provide evidence for some of the sources of validity.

The latest standards for educational and psychological testing (American Educational Research Association, American Psychological Association, &

National Council on Measurement in Education, 2014) prescribe five sources of validity evidence, that is evidence based on test content, relations to other variables, response processes, internal structure, and consequences from testing. The BAS principles especially support collecting evidence based on test content, response processes, and internal structure even during the design phases. The following subsections describe how BAS was applied to develop the CR4CR Assessment and the validity evidence collected during development where applicable.

Building Block 1: The Construct Map

The first step is to define the construct(s) that will be assessed, as has been outlined for LDC and FoI. These descriptions specify how students progress in expertise, exemplifying the first principle of a developmental perspective, in the areas of LDC and FoI. These differ from *standards* as standards do not describe development of skill or understanding and they are often presented as broad topics without a gradation in sophistication. So that we can examine the nature of the skills and knowledge within each of these topic areas, we have developed careful developmental progressions at a finer grain. We present the construct maps vertically, and as shown in the previous section, they are to be read in the order of increasing sophistication, from bottom to top. Most often, construct maps are defined as a single continuum, but we have broken from this tradition with FoI (Figure 5.4), which has the branching of the practical and statistical significance sides.

It is imperative when developing a construct map to have a clear definition of (a) what students are expected to achieve and (b) how progression toward that achievement occurs throughout a curriculum. For (a), we relied on a collection of CR and K-12 mathematics and statistics standards (described in the first section) and the expertise of an advisory committee containing math education researchers, college professors, and science and math assessment specialists. Paneling our constructs with our advisory committee of experts provided us with test content validity evidence. In terms of (b), we drew from existing literature on statistical thinking, reasoning, and literacy as described in the previous section. Following Watson and Callingham (2003), each of our constructs represents a student's tendency to ask and answer certain critical questions. It is important to note that our constructs were developed in the context of college readiness but are defined in a way to be useful for teachers before students enter college, and for instructors after they have entered. Thus, the construct maps span a range of levels, both before and after the college-ready point. Thus, we explicate the progression of a student's understanding of formal inference in FoI before, at, and beyond what might be considered college-ready. Which level in the constructs constitutes college-ready is a question that needs to be investigated with a standard-setting exercise, but this cannot be carried out until the construct map and associated assessments are in place. Moreover, it is likely that it will differ among programs of study and institutions. This is planned as the next area of research for the project. The construct maps, along with the item and outcome space materials described in the next subsections, provide the principal evidence of content validity for the instrument.

Building Block 2: The Items Design

The second principle, a match between instruction and assessment, is ensured by a principled items design. The CR4CR items were designed to produce student responses that can be mapped to the levels of the construct maps—not just at the lower, easily assessable levels but also at the higher, more complex levels of understanding. By using the construct map to guide the items design, we avoided creating a test of only basic, algorithmic knowledge, a common criticism of traditional mathematics tests. The items were designed with first-year university instruction in mind to parallel the types of thinking involved in data analysis tasks in introductory applied coursework.

To gather validity evidence based on test content and response processes we held a series of item paneling activities where we invited our advisory panel—consisting of measurement professionals, math education experts, and college instructors—to provide feedback on our constructs, each of the respective items, and how well they align to levels of the constructs. For further evidence regarding response process validity, we conducted cognitive interviews. In these interviews, respondents are asked to think aloud while they solve the assessment tasks. This is done to ensure that the thinking processes that we theorized would occur (our constructs) for a given item are actually the processes that a respondent engages in. Items were revised when we found mismatches between the respondent processes and our intended constructs and went through another round of paneling and cognitive interviews.

Both open response and multiple-choice items were developed for the CR4CR Assessment. As two main goals of the assessment design are to have rich feedback and efficient scoring, our item bank is approximately half open-ended and half multiple-choice items. Though multiple-choice items are often considered to be poor indicators of higher-level thinking skills, offering information only on rote memorizations, algorithms, test-taking skills, and other low-level knowledge and skills, our multiple-choice items are written to give information at both the low and high ends of our constructs. We do this by writing options at multiple levels of sophistication so that some choices are more sophisticated than others, but the "most correct" option is not that apparent to respondents at lower levels. These items are then scored polytomously as opposed to the traditional dichotomous scoring of multiple-choice items. Thus, none of the options is a distractor; instead each option is an attractor aligned to a particular level of the construct. This is not easily done. Qualitative data gathered in cognitive interviews and pilot test responses informed the options of the multiple-choice items so that they run the spectrum of the target construct, allowing for students to demonstrate high-level thinking (Schraw & Robinson, 2011; Ercikan & Pellegrino, 2017). For the items that survive the development process, the cognitive interviews provide information about response process validity. This type of evidence was of particular importance to us for the multiple-choice items as we wanted to ensure that they were measuring critical thinking just as well as open-ended items.

Building Block 3: The Outcome Space

The mapping of student responses to the levels of the construct map is what makes test results of the CR4CR Assessment interpretable with respect to its goal, measuring CR. This mapping constitutes the *outcome space* and is in line with the third principle, management by teachers. Sample student responses mapped to each LDC construct level for the MPG question 4 (see Figure 5.2) are shown on the right side of Figure 5.3.

Due to the complex structure of the FoI construct, we do not provide sample responses at each level in Figure 5.4. Instead, sample responses for items 1–3 are shown in the answer boxes in Figure 5.2. The response to question 1 would be scored at FoI3S,[5] the response to 2 at FoI1, and the response to 3 at FoI2.[6] Scoring guides for each item facilitate the process of categorizing student responses into the levels of the construct. An integral part of these scoring guides, which are meant to be usable by teachers, are exemplars of student performance (like the right side of Figure 5.3) that provide concrete examples of the ideal points that qualitatively differentiate the levels.

In making the test results directly interpretable in the context of the construct maps, they can be readily used by teachers to help create instructional opportunities to advance their students along the developmental continuum that is the construct map. Further, BAS stresses the importance of engaging teachers in the scoring process. For the CR4CR Assessment, we have trained some of our partner teachers to score the assessment so that they can use the results immediately instead of waiting for scores to be returned to them after external scoring.

Building Block 4: The Measurement Model

The fourth building block of BAS is the measurement model, which can provide empirical evidence of an assessment's reliability and evidence for some forms of validity. Statistical methods used for test response data include, but are not limited to, classical test theory (CTT), item response theory (IRT), latent class analysis, and factor analysis. We choose to use the Rasch family models of IRT, as this choice provides accessible interpretive tools and straightforward interpretations of student proficiency (Wilson, 2005). One of these graphical interpretive tools is the Wright Map, an example of which is provided in Figure 5.5 and which will be explained in the results section. The Wright Map is helpful in considering validity evidence for internal structure. Both students and items can be placed along the same continuum, and their locations correspond to levels of the construct map. The Wright Map can be used to verify that the item locations follow the theoretical construct map—that, for example, item scores at LDC2 are near each other and are generally above LDC1 and below LDC3. This is a very important part of internal structure validity evidence. Note that for this, it is imperative that scores are aligned to construct levels, which is why we have throughout this chapter referred to them as a construct level (e.g., FoI2) instead of a point value.

Specifically, we use a polytomous extension of the Rasch model (Rasch, 1960), called the Partial Credit Model (PCM; Masters, 1982). The PCM has parameters for each item governing the transition from each score category (i.e., each level in the construct map) to the one above it. Recall that CR4CR Assessment items were written to allow for responses at many levels of the construct, and not necessarily a simple correct/incorrect classification, so this choice makes sense. These transitions are referred to as *thresholds* and are represented graphically as the location where a student has a 50% probability of scoring at least as high as that given category. So, they can be interpreted as the point of "active learning" about the transition. A respondent who is placed, for instance, at Level 2 has not necessarily mastered that level but instead could be thought of as currently working to master that level. Note that thresholds can be transformed to any probability, but here we stick with the traditional 50% interpretation.

Results and Discussion

In this chapter, we have introduced two constructs, hence this section presents results of a two-dimensional analysis using responses from three (overlapping) forms X, Y, and Z. There were 166 respondents in total: 58 took Form X, 23 took Form Y, and 85 took Form Z. There were 29 unique items across the three forms (16 aligned to FoI and 13 to LDC), with 10 of the items acting as linking items appearing on multiple forms to allow for simultaneous estimation for the entire group of respondents.

Of the 166 respondents, 23% were high school graduates (current college students), 23% seniors, 15% juniors, 20% sophomores, and 18% freshman (1% did not provide this information) at the time they took the assessment. Because we wanted information at both low and high ends of the constructs, we recruited students in early high school through

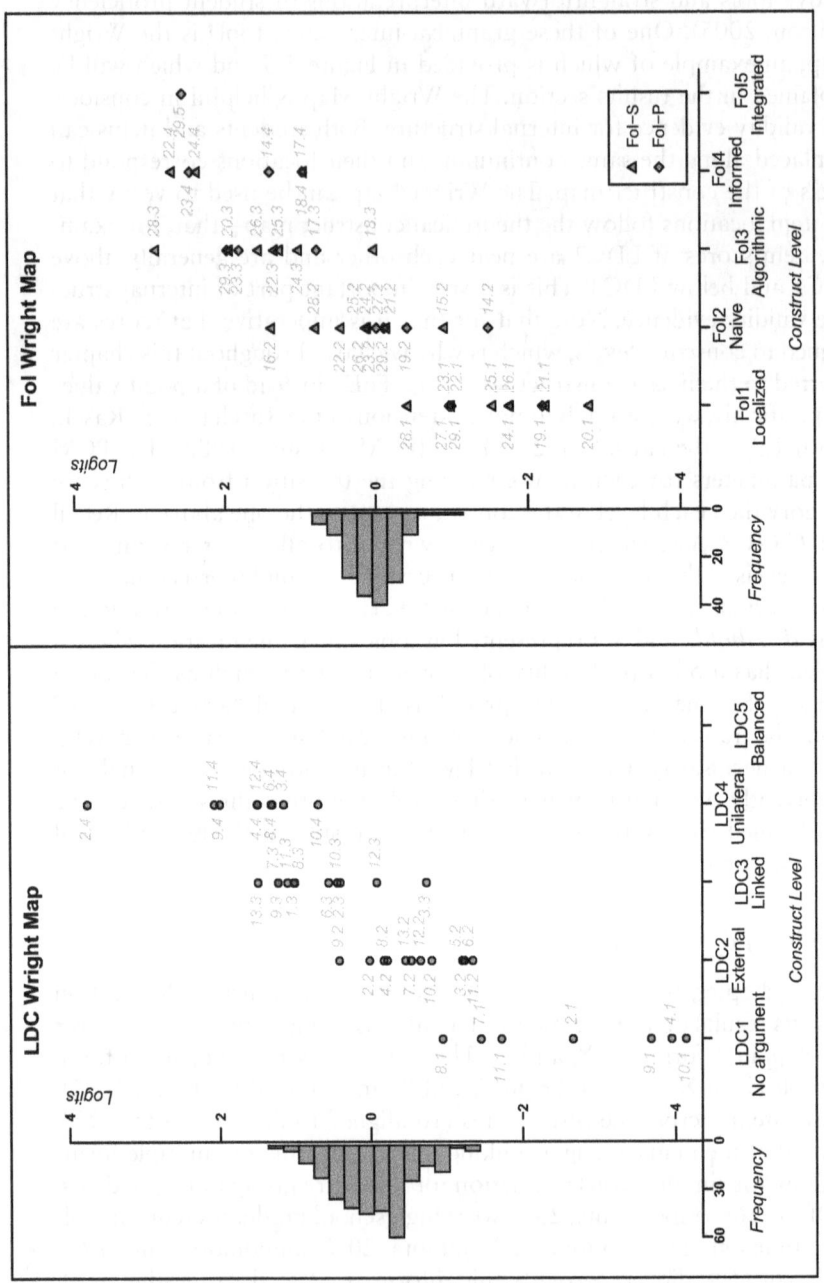

Figure 5.5 Wright Maps for the empirical analysis. Graphics produced using WrightMap package (Torres Irribarra & Freund, 2016) for R.

college. Twenty-eight percent reported having taken statistics or probability coursework in high school. The gender identification distribution of the sample was 58% female, 39% male, and 1% non-binary (2% did not provide this information).

We applied the *multidimensional approach* in which the model includes a dimension for each of the constructs LDC and FoI, and their correlation. Note that score codes for the FoI construct ranged from FoI0 to FoI5 and no distinction was made between the practical and statistical significance strands in the scoring. A score of FoI3P and FoI3S were both coded as FoI3. Note that the raw scores in the BAS framework are not considered linear but rather ordinal. That is, we stipulate that Level 4 is *above* Level 3, but not that the empirical difference between Levels 3 and 4 must be the same as the difference between any other pair of consecutive levels. We used the ConQuest software (Wu, Adams, & Wilson, 1998) to obtain model estimates.

The estimated item thresholds and person locations for the LDC construct are shown in the left panel of Figure 5.5 in a display called a Wright Map. The (on its side) histogram of person location estimates is shown on the left-hand side, with lower-scoring students at the bottom and higher-scoring students at the top. The vertical line to the right of this histogram is the *y*-axis of this map—it provides a common scale to both the person histogram and the item thresholds shown on the right-hand side—it is in logit[7] units. Along the *x*-axis (at the bottom of the map), the information differs between the left-hand and the right-hand sides. Along the left-hand side are shown frequencies for the histogram above. Along the right-hand side are the construct map labels for LDC. Above this portion of the *x*-axis, the labeled points represent the different thresholds for each item. For example, the label "8.1" at about the middle of the graph refers to the threshold associated with scoring at least LDC1 on item 8. The way that these locations are interpreted is by comparison with the location of a person in the histogram to the left. For example, a student located immediately to the left of threshold 8.1 would have an estimated probability of 50% of scoring into LDC1 or lower, a student above threshold 8.1 has more than 50% probability of that, and a person below has less than a 50% probability of that. With the Wright Maps organized in this way, if we had perfectly realized the construct map in our items, we would expect the pattern of thresholds to increase from bottom left to upper right. A perfect outcome would also show no horizontal overlap between the locations of the items in the successive construct map levels, but this is seldom perfectly realized either. As is evident in the Wright Map for LDC, we have a gradually increasing pattern as expected, which is evidence in support of the LDC construct map we wrote during Building Block 1 of BAS.

There is some overlap among the levels to be investigated. The 7.1 and 8.1 thresholds are higher than the other LDC1 thresholds and overlap.

This is interpreted as the items being affected by differences in the item prompts: LDC1 responses are characterized by a claim that is unaccompanied by an argument. The item prompts for items 2, 9, and 10 provided a claim and explicitly asked the respondent whether they agreed and then to explain. The language used in items 7 and 8 did not include an explicit prompt for a claim, thus we can expect it to be less likely that students would respond by stating a claim, and hence it would be more difficult for students to reach this level. The LDC2 threshold group is well separated from the clumps of thresholds in the adjacent construct categories.

LDC3 thresholds show reasonable grouping, though the 3.3 threshold is somewhat lower than the others. Item 3 belongs to the *Paternity Leave* (along with items 2 and 4) context, in which the stimulus material provided the contextual information needed to respond to the questions. It presented a fictional magazine that was preparing an article about their readers' support of paternity leave and a number of possible survey questions to be evaluated. This is in contrast to other item clusters, for instance the *Twitter* item cluster (items 8, 9, and 10 in Figure 5.5), in which a real-life context was presented to a student without providing the contextual information. For the *Twitter* items, respondents were presented with information about *USA Today*'s "Twitter Election Meter" which was developed during the 2012 US presidential election. Contextual information about the election was not provided. We expected, then, that the *Twitter* items would be more difficult than the *Paternity Leave* items as there is less guidance about what contextual information is relevant to the task.

LDC3 and LDC4 are less well separated than other pairs of levels, as six of the nine LDC4 thresholds are contained with the range of LDC3 thresholds. This means that there is little empirical distinction between reaching LDC3 and LDC4 for the students and items in this data set. We had very few respondents score at the LDC4 level, so it is difficult to interpret patterns or possible explanations for this overlap (due to large standard errors out at the extremes). The 2.4 threshold, for example, was determined by only one student's response, so it is not wise to speculate about why that threshold is so much higher than the others. No responses at all were registered in the highest level of the construct, LDC5. Here it is helpful to highlight the iterative nature of BAS. Recruitment for ongoing data collection is focused on ensuring we get a substantial range of abilities in order to fill out the top end of our constructs. This type of overlap between LDC3 and LDC4, should it persist after collecting a more varied sample, should also prompt more thorough investigation during Building Block 3 of BAS. When the levels are not well distinguished, it may stem from ambiguity in the scoring guide. Scoring guide decision rules based on the outcome space need to clearly distinguish responses at, for example, LDC3 and LDC4.

The right panel of Figure 5.5 shows the Wright Map for FoI. Here, we see a progression upwards from left to right and "clumps" of thresholds within the construct levels with a few items overlapping at adjacent levels, as we saw with LDC. Each FoI item was written to be targeted to either the practical (P) or statistical (S) side of the construct, and this classification is shown in the Wright Map by using different symbols: the triangles represent thresholds for S items and the diamonds for P items. This was done to see if any patterns in difficulty were apparent, for instance, if it might seem that one side of the construct was more difficult than the other. This does not seem to be the case as there are S and P items that span the ranges within each construct level. There is indication, however, of an item format effect: Items 14, 15, 17, and 18 are all multiple-choice items, and their thresholds are below the thresholds for open-ended items in FoI2, FoI3, and FoI4. Item 16, also a multiple-choice item, does not follow this pattern as its threshold is the highest in FoI2. We see a "ceiling effect" in the thresholds from levels FoI3 through FoI5.

The reliabilities[8] for LDC and FoI were 0.64 and 0.55, respectively. These are somewhat lower than what is traditionally acceptable for a standardized test, but our aim with the pilot data collection was largely to validate the proposed structure of our constructs. Further, we expect the reliability indices to improve with future larger and more variable data collection.

Conclusion

This chapter provides a demonstration for the first iteration of the BAS cycle of assessment development. Our description of the BAS cycle started when we explained our process to define a learning progression for college-ready statistical and data-based reasoning and provided the details of two constructs within that learning progression. We then described the process of item development, collected preliminary responses to our test items, and finally provided the results of the initial data analysis. The analysis results indicate that the instruments are not yet perfect—they rarely are in the first iteration—so this chapter should be used as a starting point for finer-grained definitions of college readiness in data-based reasoning and a guide for continuing development of these ideas. We are currently in our second iteration of the BAS cycle, using the analysis results presented above to further refine and improve our construct map definitions, items, and scoring guides. The construct maps presented in this chapter are the revised construct maps, and the data analyzed was produced by rescoring our pilot data using the revised versions of the constructs. So, you might consider this chapter as a showcase of our progress up through Building Block 1 of the second iteration. Our current work involves editing our existing items for clarity and for a better match to our constructs (Building Block 2), as well as adapting many of the open-ended format items to machine-scorable multiple-choice items to improve efficiency for teachers/instructors. At the same time as we revisit our items, we are refining our outcome space and scoring guides (Building Block 3) based on the responses collected in our pilot data collection as well as our ongoing cognitive interviews conducted on revised items.

Based on the validity evidence collected so far, we are confident in the structure of our proposed LDC and FoI constructs and that, while there is still much room to improve the instruments, they are indeed measuring what we intended. From the preliminary empirical results, we found that for both LDC and FoI there was some indication of banding in the increasing progression of the levels on the Wright Maps. However, there was considerable overlap of consecutive levels. This, along with the low reliability indexes of the LDC and FoI items, indicates the need for further iterations of BAS. The empirical evidence of validity may be improved by collecting a sample with more variable abilities. Collecting evidence for the other strands of validity, relationships to other variables and consequences of testing, is ongoing in our research project.

We have so far developed four other constructs within our Framework for College-Ready Data-Based Reasoning: Randomness in Study Design (RSD), Understanding Measures of Variability (UMV), Meta-Representational Competence (MRC, another extension of Lehrer et al., 2014), and Evaluating the Limitations of a Claim (ELC). Item clusters targeted at these constructs have also been developed and pilot tested in a similar manner and at the same time as those for LDC and FoI as described here. Our next steps are to explicate the relationships among the levels of the different constructs in the interest of formulating a learning progression (Center for Continuous Instructional Improvement) that goes beyond a profile of related constructs. We also have plans to develop more constructs, but these six have served as an ambitious start.

One of the most important pieces of this work will be undertaken after our first large-scale data collection, that of determining which levels of our constructs constitute college readiness. For now, we have developed the construct maps in the interest of completeness with the recognition that an entering college freshman likely does not need to be at the highest level of understanding of, for example, formal inference to be considered ready for college coursework. We anticipate that the boundaries for college readiness will differ depending on institution and degree program. First-year coursework at a large, public university may have different expectations and prerequisites than at a small, private university. Further, there may be differences in what constitutes readiness for a student entering, say, a nursing program versus an engineering program.

When the development work is concluded and we have carried out the standard setting mentioned in the previous paragraph, we will be able to release the CR4CR tests for general use. We expect that they will be used principally in two domains: the upper end of high school and the first year of college. At the former level, we expect they will be used by teachers of mathematics and statistics to help with formative assessment of individuals and classes. At the latter level, we expect they will be used to (a) help direct students into different levels of statistics classes and (b) to provide formative assessment information to teachers in statistics class.

Acknowledgments

We thank the members of our College Readiness Advisory panel who have offered invaluable feedback on our constructs and items: Jonathan D. Bostic, Richard Brown, Michele Carney, Karen Draney, Emiliano Gomez, Tzur Karelitz, Maureen Lahiff, Linda Morrell, Richard Patz, Yukie Toyama, and Danhui Zhang.

Funding

This work was supported by the National Math and Science Initiative (NMSI).

Notes

1. Both of the terms "statistical reasoning" and "data-based reasoning" are used throughout this paper. We view data-based reasoning as the more general term and statistical reasoning as a specific and important type of data-based reasoning.
2. We did not conduct a new survey of college coursework as the Conley research studies (Conley, 2005; Conley et al., 2011) cited had already done so. We used the AP courses as a second source as they represent a nationally informed source of curricular information.
3. There are 38 AP courses. We did not survey the five fine arts courses (Art History, Music Theory, 2-D Design, 3-D Design, and Drawing) as they were not expected to incorporate data-based reasoning skills in their course objectives or the associated AP exam items.
4. Our coursework and standard surveys showed that most of the statistical methods students will be exposed to are situated in the *frequentist* (conventional) framework, though Bayesian analysis is becoming more popular and it is not unlikely that students will be exposed to results produced by Bayesian methods in university coursework. Our framework is not explicitly aligned to either paradigm, though conventional p-values and confidence intervals took precedence in the assessment tasks as they are most common in introductory coursework. Good statistical reasoning supports interpretation and use of results in either paradigm.
5. The response to item 1 describes overlapping confidence intervals but is a clear misinterpretion of what a confidence interval is. Further, overlapping confidence intervals for means does *not* necessarily mean that the difference of the means is not significant. However, scoring into the Algorithmic level (FoI3S) does not require correct interpretations or attention paid to the appropriateness of the procedure.
6. The response to item 3 does discuss practical ramifications as there is some unexplained calculation of cost savings. However, it does not reach FoI3P because we cannot determine that the conclusion is based on anything other than the respondent's personal ideas about how significant 3 cents per mile is.

A response with a similar conclusion but more explanation on the calculation of the magnitude could be scored higher.
7. Logit is the term for the unit "log of the odds." It is the unit of the outcome of a logistic regression.
8. The EAP/PV reliability is a measure of test reliability or person separation reliability. It is calculated by dividing the variance of the individual EAP ability estimates by the observed person variance (Adams, 2005), and is equivalent to traditional reliability measures such as Cronbach's alpha.

References

Adams, R. J. (2005). Reliability as a measurement design effect. *Studies in Educational Evaluation, 31,* 162–172.

Agresti, A., & Finlay, B. (2009). *Statistical methods for the social sciences* (4th ed.). Upper Saddle River, NJ: Pearson Prentice Hall.

American Educational Research Association, American Psychology Association, & National Council on Measurement in Education. (2014). *Standards for educational and psychological testing* (3rd ed.). Washington, DC: Author.

American Psychological Association. (2010). *Publication manual of the American Psychological Association* (6th ed.). Washington, DC: American Psychological Association.

Biggs, J., & Collis, K. (1982). *Evaluating the quality of learning: The SOLO taxonomy.* New York: Academic Press.

Center for Continuous Instructional Improvement (CCII). (2009). *Report of the CCII panel on learning progressions in science* (CPRE Research Report). New York: Columbia University.

Conley, D. T. (2005). *College knowledge: What it really takes for students to succeed and what we can do to get them ready.* San Francisco: Jossey-Bass.

Conley, D. T. (2007). *Redefining college readiness.* Eugene, OR: Educational Policy Improvement Center.

Conley, D. T., Drummond, K., de Gonzalez, A., Rooseboom, J., & Stout, O. (2011). *Reaching the goal: The applicability and importance of the Common Core State Standards to college and career readiness.* Eugene, OR: Educational Policy Improvement Center.

Ercikan, K., & Pellegrino, J. (Eds.). (2017). *Validation of score meaning for the next generation of assessments: The use of response processes.* New York, NY: Routledge.

Evans, J. St. B. T., Newstead, S. E., & Byrne, R. M. J. (1993). *Human reasoning: The psychology of deduction.* Hillsdale, NJ: Erlbaum.

Franklin, C., Kader, G., Mewborn, D., Moreno, J., Peck, R., Perry, M., & Scheaffer, R. (2007). *Guidelines for assessment and instruction in statistics education (GAISE) report: A Pre-K-12 Curriculum Framework.* Alexandria, VA: American Statistical Association.

GAISE College Report ASA Revision Committee. (2016). *Guidelines for assessment and instruction in statistics education college report 2016.* American Statistical Association. Retrieved from www.amstat.org/education/gaise

Henderson, J. B., Osborne, J., MacPherson, A., & Szu, E. (2014). A new learning progression for student argumentation in scientific contexts. In C. P. Constantinou, N. Papadouris, & A. Hadjigeorgiou (Eds.), *E-Book proceedings of the ESERA 2013 conference: Science education research for evidence-based teaching and coherence in learning*. Part 7 (Evagorou, M. & K. Iordanou, Co-ed.) (pp. 26–42). Nicosia, Cyprus: European Science Education Research Association.

Henriques, A. (2016). Students' difficulties in understanding of confidence intervals. In D. Ben-Zvi & K. Makar (Eds.), *The teaching and learning of statistics*. Heidelberg, Germany: Springer International Publishing.

Kelley, K., & Preacher, K. J. (2012). On effect size. *Psychological Methods, 17*(2), 137–152.

Lajoie, S. P., & Romberg, T. A. (1998). Identifying an agenda for statistics instruction and assessment in K-12. In S. P. Lajoie (Ed.), *Reflections on statistics: Learning, teaching, and assessment in grades K-12* (pp. xi–xxi). Mahwah, NJ: Lawrence Erlbaum Associates, Inc.

Lehrer, R., Jones, R. S., Kim, M., Pfaff, E., & Shinohara, M. (2014). *An overview of data modeling*. Retrieved from http://modelingdata.org

Masters, G. N. (1982). A Rasch model for partial credit scoring. *Psychometrika, 47*, 49–174.

Mathematics and Statistics Standards Advisory Committee. (2007). *College Board standards for college success™: Mathematics and statistics, adapted for integrated curricula*. New York: College Board.

National Assessment of Educational Practices Governing Board. (2013). *Mathematics framework for the 2013 National Assessment of Educational Progress*. Washington, DC: US Department of Education.

National Center for Education Statistics. (2002). *NCES statistical standards* (revised ed.). Washington, DC: Department of Education.

National Council of Teachers of Mathematics (NCTM). (2000). *Principles and standards for school mathematics*. Reston, VA: The National Council of Teachers of Mathematics, Inc.

National Governors Association Center for Best Practices (NGACBP), & Council of Chief State School Officers (CCSSO). (2010). *Common core state standards: Mathematics*. Washington, DC: National Governors Association Center for Best Practices, Council of Chief State School Officers.

Newman, G. E. (2012). Bar graphs depicting averages are perceptually misinterpreted: The within-the-bar bias. *Psychonomic Bulletin & Review, 19*(4), 601–607.

NGSS Lead States. (2013). *Next generation science standards: For states, by states*. Washington, DC: The National Academies Press.

Rasch, G. (1960). *Probabilistic models for some intelligence and attainment tests*. Copenhagen, Denmark: Danish Institute for Educational Research. (Expanded edition, 1980. Chicago: University of Chicago Press).

Schaeffer, R. L., Watkins, A. E., & Landwehr, J. M. (1998). What every high-school graduate should know. In S. P. Lajoie (Ed.), *Reflections on statistics: Learning, teaching, and assessment in grades K-12* (pp. 3–32). Mahwah, NJ: Lawrence Erlbaum Associates, Inc.

Schraw, G., & Robinson, D. R. (Eds.). (2011). *Assessment of higher order thinking skills*. Charlotte, NC: Information Age Publishing.

Science Standards Advisory Committee. (2009). *College Board standards for college success™: Science*. New York: College Board.

Task Force on Reporting of Research Methods in AERA Publications. (2006). *Standards for reporting on empirical social science research in AERA publications*. Washington, DC: American Educational Research Association.

Torres Irribarra, D., & Freund, R. (2016). *WrightMap: IRT item-person map with ConQuest integration*. R package version 1.2.1. Retrieved from https://CRAN.R-project.org/package=WrightMap

Torres Irribarra, D., Freund, R., Fisher, W., & Wilson, M. (2015). Metrological traceability in education: A practical online system for measuring and managing middle school mathematics instruction. *Journal of Physics: Conference Series, 588*(1), 1–6.

von Roten, F. C. (2006). Do we need a public understanding of statistics? *Public Understanding of Science, 15*(2), 243–249.

Watson, J. (2006). *Statistical literacy at school: Growth and goals*. Mahwah, NJ: Lawrence Erlbaum Associates, Inc.

Watson, J., & Callingham, R. (2003). Statistical literacy: A complex hierarchical construct. *Statistics Education Research Journal, 2*(2), 3–46.

Wiley, A., Wyatt, J., & Camara, W. J. (2011). *The development of a multidimensional college readiness index* (Research Report 2010–3). New York: College Board.

Wilson, M. (2005). *Constructing measures*. Abingdon, UK: Routledge.

Wilson, M. (2009). Measuring progressions: Assessment structures underlying a learning progression. *Journal of Research in Science Teaching, 46*(6), 716–730.

Wilson, M., Arneson, A., Mason, J., & Wihardini, D. (2017). *Measurement of college readiness: Annual report to NMSI*. Berkeley, CA: Berkeley Evaluation and Assessment Research Center, University of California, Berkeley.

Wilson, M., & Sloane, K. (2000). From principles to practice: An embedded assessment system. *Applied Measurement in Education, 13*(2), 181–208.

Wu, M. L., Adams, R. J., & Wilson, M. (1998). *ConQuest: Generalized item response modelling software* (Computer Software Manual). Camberwell, Victoria: Australian Council for Educational Research.

6 A Validity Argument for an Undergraduate Mathematics Concept Inventory

Kathleen Melhuish and Michael D. Hicks

1. Introduction

Within undergraduate STEM (science, technology, engineering, and mathematics) education, concept inventories (CIs) have become one of the most prevalent forms of assessments. As Epstein (2013) notes, the development of CIs have become a "small cottage industry" (p. 1019). As of 2015, over 70 such assessments had been created across undergraduate STEM education (American Society for Biochemistry and Molecular Biology, 2015). The majority of these instruments share the following traits:

1. The focus is on conceptual understanding rather than procedural ability;
2. The item format is closed-form, multiple-choice;
3. Item distractors connect to genuine ways of thinking.

This emphasis on conceptual understanding is often in contrast to typical undergraduate STEM assessments focused on procedures (e.g., Tallman, Carlson, Bressoud, & Pearson, 2016).

Beyond this format and emphasis, CIs vary. Some inventories are created to be broad norm-referenced measures (e.g., Epstein, 2013), while others are created to provide careful diagnostic information connecting to levels of understanding in a theoretical framework (e.g., Carlson, Oehrtman, & Engelke, 2010). Regardless of background, the unifying validity claim for all such instruments is: The concept inventory measures conceptual understanding related to [subject]. Thus, some necessary subclaims include: (1) Each item reflects relevant conceptual understanding of [component of subject]. (2) Each item distractor connects to students' conceptual understanding [component of subject].

In this chapter, we focus primarily on one such concept inventory-style assessment: the Group Theory Concept Assessment (GTCA) (Melhuish, 2015). The chapter comprises three primary sections: a purpose section where we unpack the typical purposes for CIs and provide a detailed purpose statement for the GTCA; a background section where we unpack American Educational Research Association, American Psychological

Association, and National Council on Measurement in Education (2014) sources of validity evidence and provide a background on typical types of concept inventory validity evidence; and a GTCA validity argument section. Prior mathematics CIs have relied on reporting varying types of validation processes, often with minimal details. In this chapter we contribute a different, and we would argue more robust, approach to sharing and collecting validity evidence for mathematics CIs: a validity argument approach.

2. Concept Inventories: Their Purposes

As noted by Blaich and Wise (2011), "For assessment to be successful, it is necessary to put aside the question, '*What's the best possible knowledge?*' and instead to ask, '*Do we have good enough knowledge to try something different that might benefit our students?*'" (p. 13). In many ways, this quote exemplifies the purpose of CIs. Starting with the Force Concept Inventory (Hestenes, Wells, & Swackhamer, 1992), this style of assessment has led to direct impact and reform in classrooms via uncovering student understanding that many instructors were previously unaware, documenting how widespread particular alternate conceptions were, and providing a tool for instructors to use directly for formative assessment. In the classroom, CIs are not meant to provide students with summative grades but rather to gather information in a way that is interpretable and connected to students' ways of understanding.

2.1 The Purpose of Undergraduate Mathematics Concept Inventories

While CIs began in the undergraduate sciences, they have since moved to the mathematics domain. Besides the GTCA, research-based instruments include the Calculus Concept Inventory (CCI, Epstein, 2013), the Precalculus Concept Assessment (PCA, Carlson et al., 2010), and the Function Concept Inventory (FCI, O'Shea, Breen, & Jaworski, 2016). The purposes of these instruments are varied. The CCI, focused on conceptual understanding of differential calculus, has been leveraged to establish that reform calculus classes outperform traditional classes. The PCA, focused on graphical, computational, and covariational reasoning needed for success in calculus, is used for placement into calculus as a diagnostic tool, and as a research tool for comparing instruction type. The FCI, a measure of understanding the function concept, has uses as a potential placement tool, a formative assessment/diagnostic tool, and a research tool for comparing instruction type. Even across these three instruments with similarly leveled content, their usages vary. All three have a purpose of research evaluation tool, while the PCA and FCI also suggest additional purposes of placement and formative assessment/diagnostic tool.

The published uses for undergraduate mathematics CIs are often left to single sentences or are implicitly inferred based on uses found within publications. Standard 1.1[1] and Standard 1.3 advocate for sharing articulated purpose statements that include intended uses and interpretations as well as unintended, but likely, uses. We leverage templates developed at the Validity Evidence for Measurement in Mathematics Education[2] conference to share such a purpose statement for the GTCA.

2.2 The Group Theory Concept Assessment Purpose Statement

The GTCA measures conceptual understanding of introductory group theory, an advanced undergraduate mathematics course. Group theory is considered one of the most problematic courses for students, as it is often the first time they grapple with formal mathematics and abstractly defined concepts (Weber & Larsen, 2008). Developing conceptual understanding is essential for these students, typically mathematics majors or pre-service secondary teachers, who need this understanding to engage in productive mathematical activity (e.g., Weber & Alcock, 2004) and build connections between the course content and our base ten number system (e.g., Wasserman, 2016). We define conceptual understanding as understanding related directly to the meaning of a concept or statement (cf. Wladis, Offenholley, Licwinko, Dawes, & Lee, 2018), rather than ability to apply procedures or produce a formal proof.

In the GTCA, conceptual understanding is measured via a multiple-choice assessment where each item directly connects to a relevant concept and contains distractors linked to common ways students understand it. The assessment is meant to be used with students completing a standard undergraduate, 10-week introduction to group theory in an introductory abstract algebra course. The GTCA is available online and takes roughly 45 minutes to administer. The assessment is norm-referenced with scores measured in standard deviations. The GTCA is to be used in one of two ways: (1) as a measure at the end of an undergraduate introductory group theory course for classroom-level data (such as comparing classes under two different instructional treatments), or (2) as a formative assessment tool to collect data on students' understanding. For the second usage, the items can be given individually throughout a course. Scores are not to be used to assess individuals in any high-stakes manner, either for grades or to determine class placement. Additionally, this assessment has only been explored with students in the United States.

3. Background on Concept Inventories: Sources of Validity Evidence

As noted by Lindell, Peak, and Foster's (2007) meta-analysis, the design process and types of validity evidence collected range rather significantly

Table 6.1 Typically Collected and Shared Validity Evidence for Mathematics Concept Inventories by Source

		CCI	FCI*	PCA
Test Content	Expert Evaluation	□	□	□
	Literature Review		•	•
	Small-Scale Studies			•
Response Process	Open-Ended responses			□
	Clinical Interviews	□		•
Internal Structure	Framework Mapping		•	•
	Psychometric Item Properties		•	
	Factor Analysis	○		
	Reliability Measures	•	•	•
Relation to Other Variables	Comparing Classes	○		•
Consequences	Placement			•
	Other			

□: Conducted but no evidence provided
○: Reported, but not for validity purposes
•: Evidence provided

across different CIs, regardless of intended use. In fact, there has been substantial pushback on the validity of CIs, including the level of construct-irrelevant variance associated with the Calculus Concept Inventory (Gleason, White, Thomas, Bagley, & Rice, 2015). An overview of mathematics CIs can be found in Table 6.1.

3.1 Sources of Validity Evidence for Undergraduate Mathematics Concept Inventories

In this section, we unpack the American Educational Research Association et al.'s (2014) sources of validity evidence framework. The sources include test content, response processes, internal structure, relations to other variables, and consequences of testing. For each evidence type, we first provide a description and then appeal to the three literature-established mathematics CIs found in Section 2.1 for exemplification.

3.1.1 Test Content Evidence for Concept Inventories

Test content validity evidence supports claims that a measure represents the intended construct or domain to be measured. For mathematics CIs, this evidence tends to be of two types: expert evaluation and literature-informed item development. Both the PCA and FCI relied heavily on literature (including conducting their own small-scale studies for the PCA) to develop their items. All three instruments leveraged panels of experts (either course instructors or mathematics education researchers). For the CCI, the expert panel designed items. In contrast, the designated experts reviewing the PCA and the FCI were not directly involved with the construction of the instrument. While such evidence is often collected, there are few images of what it looks like to share this evidence.

3.1.2 Response Process Evidence for Concept Inventories

Response process evidence supports claims that the thought processes of test takers are captured by their responses to items on the instrument. Clinical interviews are the most common way to validate outcomes from response processes evidence. These interviews prompt students to talk through their process when answering an item. These interviews were conducted for the CCI, PCA, and planned for the FCI. Only the PCA provided evidence and information about the content of these interviews. Additionally, the PCA developers also collected evidence of response processes validity via their design process that began with open-ended items. The use of student responses to develop distractors (rather than expert devised) provides evidence that the distractors are reflective of student ways of reasoning.

3.1.3 Internal Structure Evidence for Concept Inventories

Internal structure evidence supports claims that the relationship among test items aligns with the nature of the construct measured. For example, analysis of dimensionality can support claims that an instrument is measuring just one thing. For mathematics CIs, the designers used a range of psychometric tools, including measures of internal consistency reliability (all three instruments) and Rasch analysis (FCI). Further, the PCA and FCI relied heavily on their underlying conceptual framework, which unpacked the targeted construct to be measured, to ground their instrument's structure.

3.1.4 Relations to Other Variables for Concept Inventories

Relations to other variables evidence supports claims that an instrument has an appropriate correlation to other measures. The developers of the aforementioned inventories provided little data in terms of relationship to other variables. This is not surprising, as CIs are designed to be distinct from existing assessments (Carlson et al., 2010). However, the PCA developers did provide evidence that students in classes aligned with the PCA's focus outscored students in traditional precalculus courses.

3.1.5 Consequence for Concept Inventories

Consequential evidence supports claims about the appropriateness of consequences from both intended and unintended uses of a measure.

Exploration of consequences is infrequently explored by concept inventory developers. The PCA did provide evidence that calculus placement is an appropriate use of the PCA via tracking students' course grades. The general lack of consequential validity evidence across the instruments is unsurprising for two reasons. First, CIs are usually developed to be low-stakes tests (with placement usage being out of the norm). The likelihood of unintended consequences is lesser. Second, it is difficult to evaluate consequences prior to adoption of the instrument to new settings. It may be that later studies could more fully address consequential validity.

This overview provides the background of current CIs. In general, the evidence collected (if not always shared) aligns with the overall purpose of CIs. Experts serve the role of evaluating that the content is appropriate for the course it is being mapped onto. Literature informs developing items connected to student conceptions. Interviews provide the grounding for interpreting student responses. Then some subset of psychometric properties is produced. We suggest that the validity evidence collected for CIs could be strengthened via attention to two issues: (1) the nature of the sample of students and experts to increase representativeness (in alignment with Standard 1.8 and 1.9) and (2) greater triangulation of sources of evidence.

4. The Content and Design of the Group Theory Concept Assessment

We employ a validity argument (Cronbach, 1988; Kane, 2004) to organize and share our triangulation of validity evidence. This type of argument has taken many forms, with emphasis usually on meaning of scores and individual item responses and consequences of scores. As noted by Schilling and Hill (2007), this focus varies on type and purpose of a given instrument. For an instrument like a concept inventory, which is low-stakes (i.e., there is no immediate consequences for an individual), the validity argument should contain claims for the meaning of scores and the meaning of individual item responses. We present such a validity argument aligned with American Educational Research Association et al.'s (2014) call for a series of claims and evidence in alignment with their sources of evidence.

4.1. The Content of Group Theory

Group theory is the study of a mathematical structure: the group. A group is a set with a binary operation that has the properties of associativity, an identity element, and inverses. Many familiar structures are groups, for example, the set of integers along with operation addition. This operation is a binary operation on the set because if one adds any two integers, the result is an integer. The operation is associative. For all integers,

$(a + b) + c = a + (b + c)$. There is an identity element (0), and all elements have inverses (for an integer a: $-a$). In group theory, students explore the group structure both within familiar contexts such as the integers and abstractly via interacting with new groups and the general group concept. Further, students explore properties of groups, relationships between groups, and varying group structures (such as quotient groups and subgroups).

To illustrate the nature of the content in group theory and on the GTCA, we share one property a group may or may not have: that of being cyclic. A group is cyclic if there exists a single element that generates the whole group. The formal definition states:

A group G is cyclic if there exists an element x such that $G = \langle x \rangle = \{x^n \mid n$ is an integer}.

Note that x^n is a general notation to reflect repeated operation. If n equals 0, then x^n is defined to be the identity element. If n is negative, x^n is defined as repeatedly operating the inverse of x. For example, the group of integers under addition is cyclic because it has a generator: 1. Any integer can be expressed by adding 1 (or its inverse, -1) enough times. The property of being cyclic is quite abstract and requires making sense of generalized notation and an abstract definition that does not fully align with the everyday usage of the word cyclic (Lajoie & Mura, 2000). The following is a GTCA item that targets this understanding:

Is $\mathbb{Z}$, the Set of Integers Under Addition, a Cyclic Group?

a. No, because $\mathbb{Z}$ is infinite and elements do not cycle.
b. No, because any element only generates part of the set (ex: 1 would only generate the positive integers).
c. Yes, because $\mathbb{Z}$ can be generated by a set of two elements.
d. Yes, because $\mathbb{Z}$ can be generated by one element.

More details about the design of this item can be found in Melhuish (2018). We use this item throughout the next sections to instantiate our validity evidence.

4.2. Details of the GTCA Design Process and Sample Selections

In order, to argue the robustness of the validity argument below, we begin by presenting an overview of the design process. In alignment with Standard 1.8, we share details of our samples to argue that they are a reasonable representation of the intended population: undergraduates in the United States completing an introductory group theory course.

The general design process of the GTCA consisted of four main stages (as seen in Figure 6.1). The first stage was a domain analysis (Messick,

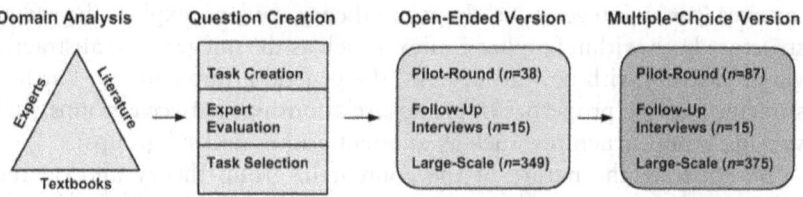

Figure 6.1 Outline of design phases for the Group Theory Concept Assessment

1995) where expert consensus determined the scope of topics covered, textbook analysis provided information about what is typical and valued in the field, and literature on student thinking provided information about alternate ways students think about concepts and theorems in group theory. From this exploration, an initial open-ended item pool was created and evaluated by experts. A subset of 18 items was then selected for testing with students. The open-ended items were piloted and field-tested with a total of 387 students (37 pilot students from four institutions and 350 field-tested students representing a stratified random sample of 24 undergraduate institutions in the United States). The 350 open-ended responses to each question were coded and categorized to become the item distractors for the closed-form version. At this point, one item was eliminated for its low difficulty. The multiple-choice version was then given to 462 students (87 in round 1; 375 in round 2) spanning 41 institutions. For this round, an attempt was made to contact all (398) current instructors of group theory across the United States via first identifying all institutions with a mathematics major, then using their course schedules to identify instructors. Roughly 6% chose to participate. The 6% of classes participating reflected a range of institution types as seen by admittance standards (See Table 6.2). During both open-ended and closed-form rounds, 15 students were interviewed for a total of 30 interviews. These interviews served to validate interpretation of student responses and probe whether their response reflected conceptual understanding related to introductory group theory. Throughout this process questions were refined based on student interviews and psychometric analysis of item properties. In addition to the main rounds, poorly performing items went through modifications and further testing. As validating is never complete, we continued to collect evidence via external users across subsequent terms.[3]

In the next sections, we outline a validity argument consisting of a series of claims and accompanying evidence within the categories of test content, student responses, internal structure, relation to other variables, and consequences. When relevant, we use the cyclic group item shared above to instantiate the types of evidence.

Table 6.2 Sample Institution Characteristics for GTCA Development Rounds

	Least Selective (>75% admitted)	Mid-Level Selective (50%–75% admitted)	More Selective (25%–50% admitted)	Most Selective (<2.5% admitted)	Not Classified	Total
Open-Ended Round	13 classes (138 students)	12 classes (108 students)	4 classes (47 students)	1 class (57 students)	0 classes	20 classes (350 students)
Closed-Form Pilot	2 classes (17 students)	0 classes	3 classes (26 students)	4 classes (44 students)	0 classes	9 classes (87 students)
Closed-Form Large-Scale	13 classes (131 students)	10 classes (127 students)	6 classes (87 students)	1 class (14 students)	2 classes (16 students)	32 classes (375 students)
Total	28 classes (286 students)	22 classes (236 students)	13 classes (157 students)	6 classes (119 students)	2 classes (19 students)	61 classes (812 students)

5. The Validity Argument for the GTCA

5.1. Claim: (Test Content Evidence) The GTCA Accurately Reflects the Domain of Introductory Group Theory

The first claim captures the degree to which a given assessment maps to a relevant content domain. In the case of the GTCA, the domain has two important aspects: the relevant setting, namely introductory group theory; and the nature of the activity captured, namely activity linked to conceptual understanding of introductory group theory. The evidence provided below attempts to address "the adequacy with which the test content represents the content domain" (American Educational Research Association et al., 2014, p. 14). As the content of the GTCA is connected to a particular course, expert instructors of that course, curricula for that course, and studies conducted on students in that course provide essential sources of evidence.

5.1.1. Subclaim: The Topics Found in GTCA Items Are Appropriately Important and Relevant to the Introductory Group Theory Level

As a field, the important content in introductory group theory is a matter of debate (e.g., Burn, 1996; Dubinsky, Dautermann, Leron, & Zazkis, 1997). Thus we leveraged experts and curricula to gather evidence for selection of topics.

PROCESS OF COLLECTING EXPERT EVIDENCE

As noted in Standard 1.11, we must justify "the procedures followed in specifying and generating test content" as well as "the domain intended to be measured" (p. 26). To begin analyzing the domain, the first step was determining which topics are relevant and important. In accordance with Standard 1.11, we explain our process of determining what is *important* in the domain. First, we conducted a Delphi Study (Dalkey & Helmer, 1963). A Delphi Study is a way of reaching consensus on an ill-defined problem. A set of experts provide responses to questions through a series of rounds. In between rounds, they receive summative information on panel responses and opportunities to provide justification for deviating from the norm. The advantage of this type of structure is to mitigate validity issues caused by other typical approaches to expert consensus, such as anonymous one-off surveys and round tables. In the case of one-off surveys, experts are not able to reflect on each other's opinions and read just based on information or ideas they may have overlooked. Round table discussions mitigate this issue but introduce a new problem: perceived status and hierarchy from the lack of anonymity. The Delphi Study mitigates these issues through a series of anonymous rounds.

Table 6.3 The Validity Argument for the Group Theory Concept Assessment

Strand and Claim	Relevant Subclaims	Nature of Evidence	Phase	Standard	Usage
Test Content Claim: The GTCA Accurately Reflects the Domain of Introductory Group Theory	Subclaim: The topics found in GTCA items are appropriately important and relevant to the introductory group theory level.	Expert Consensus Textbook Analysis	DA DA	1.11 1.11	IE
	Subclaim: The items reflect activity that is tied to conceptual understanding of group theory	Literature Review	DA	1.11	IE/ F
	Subclaim: The items minimize construct-irrelevant variance through use of appropriate symbols, vocabulary, and exemplars.	Expert Consensus Textbook Analysis Student Interviews	DA DA O/C	1.11 1.11 1.11	IE/ F
Response Process Claim: Item responses reflect students' conceptual understanding of introductory group theory	Subclaim Set: One for each item	Open-Ended Surveys Student Interviews	DA O/C	1.12 1.12	IE/ F
Internal Structure Claim: The test structure reflects structure of conceptual understanding of group theory	Subclaim: The test is unidimensional	Principle Component Analysis	C	1.13	IE
	Subclaim: The GTCA reliability is sufficient for concept inventory uses.	Cronbach's Alpha	C (EU)	1.13	IE
	Subclaim: The items cover a variety of difficulty levels all within a reasonable range and all items discriminate appropriately	Proportion Answering Item Correctly	C	1.13	IE
Internal Structure Claim: Each item distractor represents an interpretable way students think about targeted topic	Subclaim Set: One for each item	2-PL IRT model Open-Ended Surveys Student Interviews	EU O O/C	1.15	F
Relation to Other Variables Claim: The Test is Appropriately Correlated with Related Constructs	Subclaim: Scores on the GTCA moderately correlate with course grades	Spearman's Rank Correlation	C	1.18	IE/ F
	Subclaim: Students who report higher levels of learning concepts in group theory score higher on the GTCA	One-way ANOVA	EU	1.18	IE/ F
Claim: The Consequences of the GTCA Are Appropriate (No Unintended Consequences)	Subclaim: If an instructional technique or curriculum produces higher scores on the GTCA, that instructional innovation should be used more	HLM analysis, qualitative analysis of classrooms	EU	1.25	IE
	Subclaim: An instructor should adapt instruction based on formative information	Consequence unexplored			F
	Subclaim: The GTCA is not used to provide students with grades	Consequence unexplored			Other

DA: Domain Analysis; O: Open-Ended Rounds; C: Closed-Form Rounds; EU: External User Rounds; IE: Instructional Evaluation; F: Formative Assessment

For the GTCA, we adapted the methodology from the thermal and transport science concept inventory (Streveler, Olds, Miller, & Nelson, 2003). Our process consisted of four passes:

Pass 1: Experts were asked to compile a list of concepts they think are essential in introductory group theory.

Pass 2: A list was compiled of all concepts mentioned by at least two experts. The experts then rated each topic on a scale from 0 to 10 for difficulty and importance.

Pass 3: The experts were provided with the 25th, 50th, and 75th percentile scores for both categories and asked to rate again. During this pass, the experts provided justifications for any rating outside of the 25th–75th percentile range.

Pass 4: Experts were provided with the same numerical information as well as the justifications from pass 3, and were asked to provide a final rating.

In accordance with Standard 1.9, we share details of our expert panel, arguing it comprises experts with the relevant experience. The GTCA expert panel contained experienced group theory course instructors (taught at least three times, average 10). The panel was created to be heterogeneous. The 11 panelists who provided final ratings included four group theory textbook authors, eight mathematicians (split between group theorists and other disciplines), and five mathematics education researchers with publications related to the teaching and learning of group theory. These categories are not mutually exclusive.

EXPERT EVIDENCE

Through this process, the experts arrived at consensus regarding the standard topics of an introductory group theory course. All topics covered in the GTCA had a mean importance rating of at least 9.0 out of 10.0. (The average importance score for evaluated topics was 6.83 (SD = 2.57), meaning these topics were at least 0.85 standard deviations above the average.

SAMPLE ITEM CLAIM: CYCLIC GROUP IS AN IMPORTANT TOPIC IN
INTRODUCTORY GROUP THEORY

The topic most related to the sample item is cyclic groups. The average importance rating was 9.55, roughly one standard deviation above average rating (with all panelists rating the topic as 9 or 10).

PROCESS OF COLLECTING TEXTBOOK EVIDENCE

The second type of evidence for the importance of topics came from a textbook analysis. As part of the intended curriculum (Travers & Westbury,

1989), textbooks serve as a proxy for what is treated in class. We first identified the most commonly used textbooks via surveying a random sample of institutions in the United States that offered a mathematics major. We then identified which textbook was used for their introductory undergraduate abstract algebra (or group theory) course via online bookstores or contacting the most recent instructor. We surveyed until we reached 294 responses, allowing for a 95% confidence interval with error of ±5%. We included all textbooks used by 5% or more of schools: Fraleigh (2003), Gallian (2013), Gilbert and Gilbert (2009), and Hungerford (2013).

TEXTBOOK EVIDENCE

All topics identified by the panel were treated heavily in early sections of the textbooks. Across the texts, these topics were the focus of at least 70% of the textbook chapters and/or sections serving as the introductory group theory unit.

SAMPLE ITEM CLAIM: CYCLIC GROUP IS AN IMPORTANT TOPIC IN
INTRODUCTORY GROUP THEORY

Cyclic groups were introduced within the first four group theory chapters/sections across all four textbooks. In three of the four textbooks, they had their own dedicated section. In the fourth book, they constituted roughly half of one section.

COUNTEREVIDENCE, LIMITATIONS, AND ALTERNATE INTERPRETATIONS

While the evidence largely pointed to the importance of these concepts across the subject area, it is important to acknowledge limitations. First, there were only four textbooks with a sizeable portion of the market share. However, there was a total of 41 textbooks in use. It is possible that the treatment of topics in those books may be inconsistent. Further, the expert panel is a sample of instructors particularly dedicated to teaching this course. It may be that other instructors, those who may not opt into an unpaid study, have other values. Additionally, more demographics could strengthen this evidence.

5.1.2 Subclaim: The Items Reflect Activity That Is Tied to Conceptual Understanding of Group Theory

The test content is meant to cover a certain type of knowledge of group theory: knowledge connected to conceptual understanding. As such, the items in the GTCA need to be connected to conceptual understanding.

PROCESS OF COLLECTING LITERATURE EVIDENCE

One source of distinguishing between tasks connected to conceptual understanding and tasks that would convey other types of knowledge is by identifying task types documented within the literature. We conducted a thorough literature review for each targeted topic to identify: documented issues of conceptual understanding, and tasks and activity types associated with probing students conceptual understanding. For the GTCA, we define rich, conceptual understanding as understanding that is internally coherent and rich in relationships (e.g., Hiebert & Lefevre, 1986); and understanding that is connected directly to (or logically necessitated by) the definition (or meaning) of a concept or mathematical statement (cf. Wladis et al., in press). In accordance with Standard 1.11, we attended to the appropriate "cognitive complexity" (p. 26) to develop tasks that mimic activity types documented to connect to student understanding from the literature.

LITERATURE EVIDENCE

We claim that each of our tasks meet these criteria and are aligned with activities that can unearth conceptual understanding according to prior studies. Each item in the GTCA corresponds to one of the following activities: navigating between formal and informal definitions (e.g., Lajoie & Mura, 2000); navigating between the process of creating an object and treating the object itself (e.g., Dubinsky, Dautermann, Leron, & Zazkis, 1994); applying a definition to determine if an example meets the criteria (Novotná & Hoch, 2008); generating examples meeting certain criteria (e.g., Zaslavsky & Peled, 1996); and knowing what contexts align with theorem application (e.g., Hazzan & Leron, 1996).

However, these types of activities can be productive or unproductive in unveiling students' conceptual understanding, depending on whether a context is sufficiently abstract (Oktaç, 2016). If the context is not abstract, a memorized rule may supplant conceptual reasoning. At the task level, all tasks for the GTCA items have contexts that are sufficiently abstract (in terms of Hazzan's (1999) classification) to allow for conceptual understanding to be guiding question response.

SAMPLE ITEM CLAIM: ACTIVITY INVOLVED IN DETERMINING IF Z IS
CYCLIC IS CONNECTED TO CONCEPTUAL UNDERSTANDING OF CYCLIC

The task in the sample item was to determine if an example group, the group of integers, met the definition for being cyclic. This is naturally connected to understanding the definition of cyclic. In this case, Lajoie and Mura (2000) previously documented that the infinite group Z is problematic for students who have not fully conceptualized the formal definition and connected it with their intuition.

PROCESS OF COLLECTING EXPERT EVIDENCE

Each item from the original open-ended pool of 42 items was evaluated by at least two mathematicians who have experience teaching group theory and at least two mathematics educators who have published related to group theory pedagogy. For all questions, they were asked to evaluate:

1. Is this task relevant to an introductory group theory course?
2. Does this task represent an important aspect of understanding [relevant topic]? (The task might be too advanced, capture prerequisite knowledge, or be unrelated to the targeted concept.)
3. What student conceptions might this task capture? (You could speak generally, or provide samples of how you think students might respond.)

EXPERT EVIDENCE

The original set of items in the GTCA was selected based on unanimous agreement on questions 1 and 2, reflecting both relevance and connection to understanding.[4]

SAMPLE ITEM CLAIM: DETERMINING IF Z IS CYCLIC IS A RELEVANT
AND IMPORTANT TASK IN INTRODUCTORY GROUP THEORY

For the sample item, three reviewers identified the item as both important and relevant. This was an anomaly, as the fourth reviewer did not answer the first two questions, providing only an open-ended response. However, this question stemmed from the literature and thus, paired with the agreement across the responders, was selected.

COUNTEREVIDENCE, LIMITATIONS, AND ALTERNATE INTERPRETATION

While the items were designed to align with conceptual understanding of important topics in group theory, this was not always the case, and in fact when piloted, two questions allowed for students to bypass their conceptual understanding and defer to a rote procedure. While the items were designed to connect to the literature, the items did not always perform as anticipated and thus additional evidence from field-testing is needed to more fully argue the case for conceptual understanding. (See Section 5.2.)

5.1.3 Subclaim: The Items Minimize Construct-Irrelevant Variance Through Use of Appropriate Symbols, Vocabulary, and Exemplars

One of the biggest places where content validity of a test comes into play is the minimization of construct-irrelevant variation (Messick, 1995). That is, students' performance on the test should not be dependent on any attribute beyond their conceptual understanding of group theory. Things such as access to language or knowledge of a context can interfere with the ability to answer a prompt. Standard 1.11 refers to this as "the accessibility of the test content to all members of the intended population" (p. 26). For example, when mathematics prompts are set in real-world settings, they may be biased toward those with relevant experience. In the case of group theory, many of these issues are avoided by the nature of the subject area. There are no real-world contexts that could impact access to the prompts. However, there are mathematical issues that can arise. The goal of this assessment is to be useful regardless of which curriculum students use. In the prior claims, we established that the content covered was universal. That is, we did not advantage test takers by choosing a set of topics that may only be in certain curricula. We now attend to other issues of access: which examples, representations, and notation are universal.

PROCESS OF COLLECTING EXPERT EVIDENCE

To determine which examples were universal enough to provide the context to ask meaningful questions, we return to the Delphi Study. The experts were asked to identify the essential topics of group theory. On this list were many example group contexts. The experts diverged on the necessity of these groups in an introduction course. Therefore, a reasonable assumption was that not all examples would be accessible to all students with variance in treatment by instructor.

EVIDENCE

The examples, or contexts, in the GTCA items are only example groups where the experts reached agreement (modular arithmetic groups) or groups from the K-12 curriculum (such as the integers under addition).

PROCESS OF COLLECTING TEXTBOOK EVIDENCE

In addition to verifying topics, we conducted a thorough textbook analysis (e.g., Mesa, 2004) of each curriculum's treatment of the essential topics. This analysis included identifying the prevalence of various representation types, the similarities and differences across formal and informal definitions, the nature and type of examples, and the purposed or expected study activity.

TEXTBOOK EVIDENCE

All items in the GTCA leverage examples that are treated heavily across all textbooks, representations (and notation) that are universal, and tasks that are meaningful regardless of definition treatment. All example groups in the GTCA are either modular groups (found in at least 75% of sections across textbooks) or groups familiar from the K-12 curriculum. This analysis prevented us from using one of the most common example groups in the literature, the dihedral group (e.g., Dubinsky et al., 1994). In some books, the dihedral group served the role to motivate the definition of group and was used repeatedly for many example purposes. However, in other texts, the dihedral group was mentioned briefly (with a non-standard name) and hardly leveraged. For this reason, despite its prevalence in research literature, no GTCA questions use the dihedral group.

PROCESS OF COLLECTING INTERVIEW EVIDENCE

The last piece of evidence stems from student interviews. For both the open-ended items and close-formed items, we interviewed sets of 15 students (for a total of 30^5 students) to have them explain their thinking processes for their responses. During the interview process, we verified that students understood all notation, example groups, and terminology.

INTERVIEW EVIDENCE

Across all interviews, students were able to make sense of all notation, were familiar with all example groups, and did not encounter unfamiliar terminology with one exception. One type of group, modular arithmetic groups, has elements that are symbolized in four different ways depending on textbook and instructor (a, $[a]$, $[a]_n$, $\bar{a}$). During interviews, it became clear that the initial notation used, $[a]_n$, was not accessible for students who learned the a notation. After testing different notations,[6] only $\bar{a}$ was understood by all students. Thus, this notation is used in the GTCA items.

SAMPLE ITEM CLAIM: THE EXAMPLE CONTEXT, THE INTEGERS, AND REPRESENTATION, VERBAL AND SYMBOLIC, IS ACCESSIBLE TO STUDENTS IN INTRODUCTORY GROUP THEORY

In the cyclic group item, the example group is $\mathbb{Z}$, a notation found in all texts. Additionally, we provided the meaning, "integers under addition" to mitigate for any ambiguity by using the symbol alone. This group was used consistently across the four analyzed curricula, serving many roles, which include being an example of a cyclic group in all four texts. Additionally, this is a group that students are familiar with from their K-12 education.

COUNTEREVIDENCE, LIMITATIONS, AND ALTERNATE INTERPRETATION

While we largely argue for the accessibility of items for the targeted population, we acknowledge that we may have failed to detect issues of wording or examples with only 30 interviews. Further, one of the 30 interviewed students indicated that he spent little time on the modular arithmetic groups (a standard example across the measure.) It is possible that all students may have access to this example context, but their depth and comfort level may vary according the emphasis in their particular class.

5.2 Claim: [Response Processes] Item Responses Reflect Students' Conceptual Understanding of Introductory Group Theory

In CIs, this strand of evidence is the most essential. Every item is designed to be interpretable and every distractor of every item should have meaning related to students' conceptual understanding. As outlined in Standard 1.12, evidence is needed for the "cognitive operations of test takers" (p. 26). Methodologies such as cognitive interviews where students think aloud can provide such evidence.

Process of Collecting Open Response Evidence

The first type of evidence comes from the design process of the GTCA. Rather than begin with closed-form items generated from researcher's views of student understanding, the items all began open-ended with over 350 students' responses collected for each time. In this way, students' ways of thinking could be documented without influence of a distractor set.

Open Response Evidence

For the items in the GTCA, all distractors stem from students' open-ended survey responses and thus represent genuine ways that students think about group theory content.

Sample Item Claim: Each Response to the Multiple-Choice Question Reflects Reasoning Related to Conceptual Understanding of Cyclic Groups

The multiple-choice distractors from the sample item stem from student open-ended responses. See Table 6.4 for open-ended responses and the related category type.

Process of Collecting Interview Evidence

The second type of evidence comes again from conducting think-aloud interviews with students. In addition to checking issues of accessibility, we also used these interviews to probe students' reasoning for responses. In each case, students were asked to explain their thinking for their response. Follow-up questions served to further probe their thinking. If

Table 6.4 Open-Ended Responses for Cyclic Question

Response Category	Sample Student Response
$\mathbb{Z}$ is cyclic because it can be generated by one element	Yes. To be cyclic means that a single element of $\mathbb{Z}$ can produce the entire set, which is $\mathbb{Z}$ under the following condition: $\langle a \rangle = a^i$ (or some integer $i \in \mathbb{Z}$, Letting $a = 1$ will produce all of $\mathbb{Z}$.
$\mathbb{Z}$ is cyclic because it can be generated by a finite set of elements	Yes, we can start w/ $1 \& -1$ and generate all of $\mathbb{Z}$
$\mathbb{Z}$ is not cyclic because the elements do not cycle	No. If you take any number and keep applying it to itself (addition) you will never get back to that number while having gotten every other number since $\mathbb{Z}$ is infinite.
$\mathbb{Z}$ is not cyclic because no element will generate the whole set	No. Cyclic means generated by an element. If you start with a negative element, it can only generate a set of negatives. If you start with a positive element, it can only generate a positive set. So no element can generate $\mathbb{Z}$.

their response did not explicate their understanding of a topic, they were prompted to directly share their understanding of a given topic to assure alignment with their reasoning on the task.

Interview Evidence

For all but one item, the interviews verified that students provided responses on the GTCA items for the hypothesized reasons. As these reasons vary per item, we share in-depth evidence for our sample item.

Sample Item Claim: Students Selecting Each Response Do So for Reasons Aligned With Their Conceptual Understanding of the Integers

We illustrate this type of evidence by sharing representative pieces of interviews from students selecting each of the response options for our sample item. Note that these multiple-choice responses directly connect to the open-ended responses from Table 6.4. Students who selected the correct response ("Yes, because $\mathbb{Z}$ can be generated by one element") voiced comfort with n taking on negative values and/or using an element and its inverse when thinking about the set generated by a given element. For example, one student explained, "It's generated in the way you take powers of one or in this case, you take one plus one plus one and inverses and you get the whole set back." When asked why inverses are also used, he added, "You take n powers of all elements in and n can include negative powers and inverses." This response connects clearly to the meaning of generating and cyclic.

A student selecting the response "Yes, because $\mathbb{Z}$ can be generated by a set of two elements" reflects a conflation between a single generator and a set of generators as the basis for a group being cyclic. One student explained, "it was cyclic because if you have the generators one and negative one the whole group is generated." This response reflects that the student has not conceptualized important definition aspects: (1) using the element one generates positive and negative integers, and (2) cyclic groups need one generator and not a set of generators. When further probed about whether it was always sufficient to find two elements that generate or any finite set of generating elements, the student responded, "I don't know if there is any significance to those specific numbers." Further, he went on to explain that he did not know if there were any limitations on the number of elements in a generating set to meet the definition of cyclic.

A student selecting the response "No, because $\mathbb{Z}$ is infinite and elements do not cycle" reflects a general image of groups being built by repeatedly operating with a single element until it cycles back. One student explained:

So, if the order is infinite that means—give me a second—alright, so if you have a thing that does has infinite order, you get there's infinite, once you get that you'd cycle back to the beginning of that set—that's how you'd know there was a generator, but since there is an infinite group, it never cycles back to it.

This response reflects the focus on the word "cycle" colloquially, where a cyclic group must "cycle back," which cannot happen if elements continue infinitely. However, the definition of cyclic does not require elements to cycle in such a case.

A student selecting the response "No, because any element only generates part of the set (ex: 1 would only generate the positive integers)" is focusing on generating as repeated operation and recognizes that a single element must generate the set. However, they do not attend to essential role of negative powers. One student explained:

So, if you take any positive integer and operate it under addition, you are only ever going to get positive integers. And similarly, for a negative integer. Zero won't generate any integers. You are never going to have a single element that generates the entire cyclic group.

When prompted, she verified her definition for generator as "one element that with repeated composition with itself gives all elements of the group." This definition does not align with the formal definition that allows for exponents that are positive, negative, and 0.

Despite $\mathbb{Z}$ being a standard cyclic group in the group theory curricula, many students demonstrated that their conceptual connection to cyclic bypassed the definition (and likely their prior exposure to this example).

Counterevidence, Limitations, and Alternate Interpretations

There was one item that performed poorly on this front: an item about isomorphism. In the case of this item, students were able to arrive at the correct response using a procedural approach without conceptual understanding of isomorphism. A more in-depth discussion of this item and problematic approaches can be found in Melhuish (2018). Ultimately this item was altered and refined through additional open-ended testing. However, interview data still needs to be collected.

5.3 Claim: [Internal Structure] The Test Is Unidimensional; All Items Have Appropriate Difficulty and Discrimination Characteristics

The intention of the GTCA was not to capture various dimensions of conceptual understanding. Rather the intention was to build a set of items that all correspond to conceptual understanding of group theory but provide individual information related to particular topics.

5.3.1 Subclaim: The Test Is Unidimensional

The overall structure of the GTCA was designed to be unidimensional. That is, there are not multiple factors that account for performance.

PROCESS OF COLLECTING DIMENSIONALITY EVIDENCE

As suggested in Standard 1.13, we analyzed dimensionality to address this concern. We used a principle component analysis via SPSS. We determined the dimensionality via exploring a scree plot (Williams, Onsman, & Brown, 2010).

DIMENSIONALITY EVIDENCE

Figure 6.2 represents the resulting scree plot (extracted components plotted against their eigenvalues). There is a noticeable elbow after the first

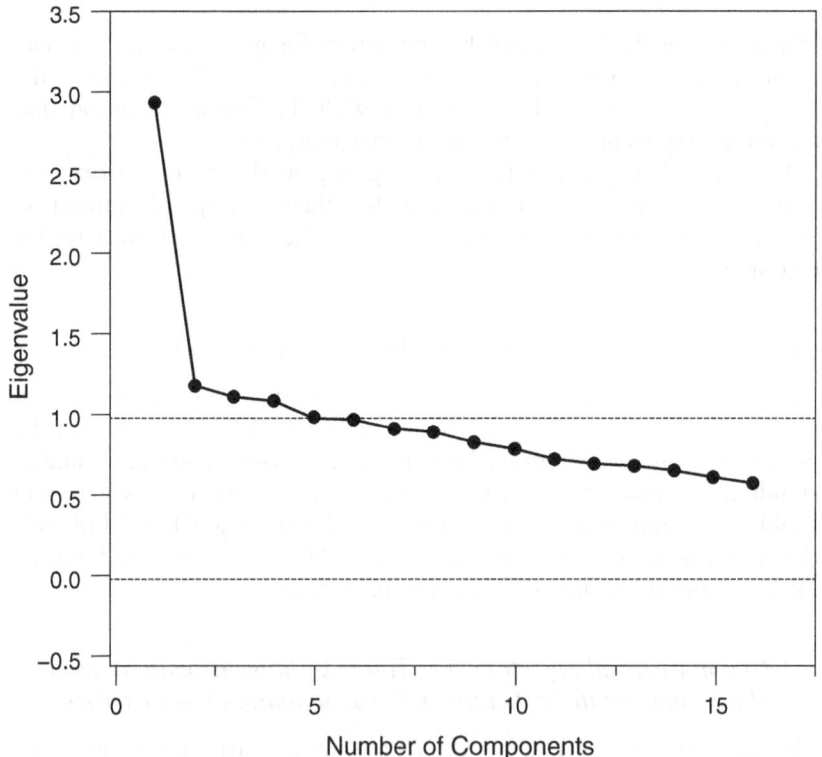

Figure 6.2 Scree plot from GTCA analysis

component. From this plot, it is reasonable to treat the GTCA as unidimensional. Thus, "the variability attributed to one major dimension was much greater than the score variability attributable to any other identified dimension" (Standard 1.13, p. 27).

COUNTEREVIDENCE, LIMITATIONS, AND ALTERNATE INTERPRETATION

A standard dimensionality analysis uses binary data, right or wrong, for each item. However, for CIs, each item distractor has meaning. Thus, such an analysis may not be detecting underlying structure that might account for the selection of particular responses.

5.3.2 Subclaim: The GTCA Reliability Is Sufficient for Concept Inventory Uses

PROCESS OF COLLECTING RELIABILITY EVIDENCE

Cronbach's alpha was calculated through all rounds of closed-form GTCA testing.

RELIABILITY EVIDENCE

Alpha ranged from 0.65 to 0.84. For the latest administration of the stable version of the GTCA, $\alpha = 0.73$. This is generally deemed an appropriate level of reliability for a low-stakes test such as a concept inventory where items are dichotomous (Epstein, 2013).

COUNTEREVIDENCE, LIMITATIONS, AND ALTERNATE INTERPRETATION

While this alpha is acceptable, it could be higher. There are limitations, as noted above, at looking at a rough "right" or "wrong" answer and determining reliability from there. Additional reliability analysis, such as parallel tests or more advanced item response theory methods, could strengthen reliability claims.

5.3.3 Subclaim: The Items Cover a Variety of Difficulty Levels All Within a Reasonable Range and All Items Discriminate Appropriately

PROCESS OF COLLECTING ITEM CHARACTERISTIC EVIDENCE

This claim was addressed throughout the design process. During early stages, all items were correlated to overall score using a point-biserial correlation. Item difficulties were calculated via percentage of students answering correctly. After the largest round ($n = 349$), a 2-parameter item response model was created to better explore item characteristics (Baker, 2001). A

2-parameter item response model identifies an item difficulty and item discrimination for each item via fitting item data to a logistic model. This information corresponds to what ability level was needed to have a 50% chance of answering the item correctly (item difficulty), and the ability of an item to sort between those above and below that 50% line (item discrimination).

ITEM CHARACTERISTIC EVIDENCE

During early stages, all items were correlated positively with overall score and with a range of 18% to 70% of students answering each item correctly with one exception: a question targeting isomorphism which was only answered correctly by 12% of students. Further, the 2-parameter item response model verified this difficulty problem. The isomorphism topic had a difficulty of 14.1214 and item discrimination of 0.1422. This reflects that a student would need an ability level of 14.1214 standard deviations above the mean to have a 50% chance of answering this question correctly. This is obviously a nonsensical breaking point. Thus the item was sufficiently altered to reduce its complexity. With this substitution, the new item is still quite difficult (2.48), but more reasonable. All other items in the GTCA range in difficulty from –1.01 to 1.54. All items positively discriminated with no items at very low. Three items were in the low discrimination range (0.35–0.64), seven items were moderately discriminating (0.65–1.34), and seven items were highly discriminating (1.35 and higher).[7]

SAMPLE ITEM CLAIM: THE CYCLIC GROUP QUESTION WAS NOT TOO DIFFICULT OR TOO EASY; THE CYCLIC GROUP QUESTION POSITIVELY DISCRIMINATED BETWEEN STUDENTS SCORING BELOW AND ABOVE 0.98 STANDARD DEVIATION ABOVE AVERAGE

The sample item had a difficulty of 0.98 reflecting that a student scoring roughly one standard deviation above average had a 50% chance of answering the item correctly. The item had a moderate discrimination: 0.50. Note that this was the least discriminating item on the GTCA.

COUNTEREVIDENCE, LIMITATIONS, AND ALTERNATE INTERPRETATIONS

First, the new isomorphism item's difficulty is 2.48, which is still a bit higher than ideal. As such, this item should be used with caution. Additionally, the easiest items on the test still have a difficulty level of roughly –1.01. That means that no items have a 50% of correctness for anyone roughly below one standard deviation from the mean. A more optimal range of items might include a –1.5 or even –2.0 difficulty item to better discriminate among low scores. We conjecture that the difficulty of abstract algebra may make the development of particularly easy items non-attainable. Additionally, the three low discriminating items could be improved.

5.4. Claim (Relation to Other Variables) the Test Is Appropriately Correlated With Related Constructs

This type of validity claim is less essential for CIs (Carlson et al., 2010). By their nature they are measuring something that is distinct from other assessments. However, there are some reasonable relationships that can be verified. In accordance with Standard 1.16, we explain the rationale for relating these constructs. Further, in accordance with Standard 1.20, we provide measures of significance and effect size.

5.4.1. Subclaim: Scores on the GTCA Moderately Correlate With Course Grades

Course grades should be positively related to GTCA scores but not so related that measure is redundant.

PROCESS OF COLLECTING GRADE CORRELATION EVIDENCE

Course grades typically rely heavily on formal proof production (which is informed by but not equivalent to conceptual understanding). To test this relationship, we used a Spearman's rank correlation analysis on the largest round of GTCA respondents who self-reported grades (n = 365). We treated grades as ordinal data on a standard scale (0–4).

GRADE CORRELATION EVIDENCE

Course grades were positively correlated to GTCA scores, ρ = .4306, and significant, $p < .001$. The correlation coefficient reflects a medium effect size (Cohen, 1992).

COUNTEREVIDENCE, LIMITATIONS, AND ALTERNATE INTERPRETATIONS

The correlation with course grades could be strengthened via more robust analysis, such as leveraging a hierarchical model to account for the correlation between students in classes. Further, evidence concerning grade point average or math SAT scores should be collected to explore the degree to which this correlation reflects broader mathematical or educational ability.

5.4.2. Subclaim: Students Who Report Higher Levels of Learning Concepts in Group Theory Score Higher on the GTCA[8]

It is reasonable to assume there should be some correlation between students' self-reported degree of understanding concepts in group theory and their GTCA scores.

PROCESS OF COLLECTING REPORTED LEARNING CORRELATION EVIDENCE

During a recent external use of the GTCA, a research team collected data on students' self-reported gains in understanding in group theory. Ninety-five students reported little, moderate, good, or great gains in understanding. We note that these students were all part of experimental inquiry-oriented abstract algebra classes and are not necessarily representative of the greater population. A one-way analysis of variance (ANOVA) was conducted to determine if there were differences in GTCA scores across the condensed categories of low gains (little to moderate), medium gains (good), or high gains (great).

REPORTED LEARNING CORRELATION EVIDENCE

There was a significant relationship between perceived gains in understanding of concepts and GTCA scores for the three levels [$F(2, 94)$ = 3.36, p = 0.04]. Post hoc comparisons using the Tukey HSD test indicated that the GTCA score for the high gains group (M = 0.25, SD = 0.95) was significantly higher than the low gains group (M = -0.36, SD = 0.73). This difference reflects a moderate effect size, d = .71. However, the medium gains group did not significantly differ from either (M = -0.15, 1.08). This analysis reflects that GTCA scores do reflect the difference between students who identified low gains in their understanding of concepts from the course and those who identified high gains in their understanding of the course.

COUNTEREVIDENCE, LIMITATIONS, AND ALTERNATE INTERPRETATION

Self-reports of understanding are a dubious construct. As such, this evidence is promising for differentiating students who have confidence in their understanding gains; however, this can only support minimal claims.

5.5. Claim: The GTCA Does Not Have Unintended Consequences

One way to think about consequential claims is to think about intended and unintended consequences in alignment with the purpose statement. The GTCA is meant as a form of formative assessment and research tool, but it is not meant to provide summative assessment of individual students. In that way, unintended consequences are unlikely, such as "teaching to the test," which might occur with assessments with more dire consequences. Additionally, there are no pass/fail scores that would have implications for students in terms of their grades and continued mathematics careers.

5.5.1 Subclaim: The GTCA Is Not Being Used to Assign Students Grades

The GTCA is currently only available online behind a password-protected wall. Data from the GTCA is given at the aggregate level. However, once an instructor accesses the instrument, they could adapt it for their own usage, which could lead to unintended consequences if used to assign grades.

PROCESS FOR COLLECTING USAGE EVIDENCE

Further research would be needed to see if the GTCA was used beyond its intended usage. Once the GTCA is in more wide-scale usage, instructors could be surveyed about whether, and how, they are using any of the items in class and for what purposes. Further, if the GTCA is being used against its purpose, additional exploration of the consequences of grade impact would be needed.

5.5.2 Subclaim: Instructional Innovations Should Be Continued or Ceased Based on Evidence for the GTCA

The consequences of intended use could also be explored in future studies. A consequence may be the increased use of an instructional innovation if students are documented to develop increased conceptual understanding under that treatment. This is a valid consequence if the scores are meaningful and connected to what we want students to learn in group theory classes. However, as described in Standard 1.15, group differences must lead to a re-exploration of the test itself to determine if performance is not impacted by construct-irrelevant variance.

PROCESS FOR COLLECTING GROUP DIFFERENCE EVIDENCE

Concept inventories are often used to compare classrooms and study the impact of pedagogical approaches. To an extent, this consequence is valid via the strength of the argument above. However, additional evidence should be collected to assure that difference in performance on the GTCA is not tied to limited treatment of certain topics in a particular curriculum. Further, this consequence reflects that difference in performance is tied entirely to the targeted treatment. However, instructors vary across any numbers of dimensions. For example, research has established that instructors using an inquiry-oriented curriculum created a gender gap in performance on the GTCA (Andrews-Larson, Can, & Angstadt, in press). An unintended consequence would be the abandonment of inquiry-oriented instruction without the collection of additional validity evidence for the GTCA, and additional exploration of treatment conditions.

6. Discussion

A validity argument serves to produce a series of claims for interpretation and use of a given instrument and substantiating evidence. We advocate that the validity argument may be a particularly powerful approach to guide collecting and presenting of validity evidence for mathematics CIs. The GTCA validity argument provided a series of claims and evidence that may be appropriate to argue for a CI's dual purposes: research and formative assessment. In general, the evidence compiled served to argue that the content of the test (i.e., the range of topics and types of items) accurately reflected important areas in group theory. Thus, it would be reasonable to use this measure to compare classes because the GTCA was designed to be accessible to students regardless of their curriculum or instruction. Evidence was collected based on a heterogeneous group of experts and representative samples of students. Additionally, this instrument can serve a formative assessment purpose due to the meaningfulness behind every distractor. By starting with open-ended responses, converting to closed-form items, and interviewing students across rounds, we collected evidence that students engaged in the hypothesized thought processes documented various ways of understanding important topics. We argue that test content and response process evidence is the most essential for CIs' purposes. However, the validity argument is strengthened through the addition of internal structure and relation to other variables evidence. To make a more robust argument, next steps would include collecting consequence evidence regarding the GTCA's usage and empirically exploring the consequences of use. Further, additional analysis and data collection in all categories would strengthen our argument, particularly as items are refined in light of continued usage. We do not mean to convey that the present validity argument is the final product, rather that collecting validity evidence is an ongoing process.

We conclude by reflecting on validity evidence driving the design process of the GTCA. We conducted a domain analysis prior to its creation. This analysis included a systematic approach to expert consensus, substantial literature review, and analysis of related textbooks. This process provided evidence related to test content validity in terms of the importance of topics included, the nature of tasks created, and the claims of minimizing construct-irrelevant variance. The decision to follow Carlson et al.'s (2010) methodology for the creation of the closed-form items rather than research-created distractors produced substantial evidence that the item distractors connected to ways students think about the relevant concepts. Each stage of development partnered with the collection of validity evidence. As such, we challenge the paradigm outlined in Lindell et al.'s (2007) meta-analysis of concept inventory design: instruments are first created, then validity studies are conducted. Rather, the validity argument should guide all design decisions during measurement creation.

Notes

1. Throughout this chapter, "standards" refers to the American Educational Research Association, American Psychological Association, & National Council on Measurement in Education's (2014) *Standards for Validity*.
2. The *Validity Evidence for Measurement in Mathematics Education* was organized and funded through a National Science Foundation grant (DRL-1644314, 1644321). Any opinions, findings, and conclusions or recommendations expressed in this material are those of the author(s) and do not necessarily reflect the views of the National Science Foundation.
3. External user round: 17 classes, 191 students.
4. The current version of the GTCA contains 15 of 17 items that are close adaptations from items evaluated by the expert panel.
5. Carlson et al. (2010) recommend interviewing 10 students per cycle. We expanded to 15 to ensure we interviewed students selecting each of the distractor options.
6. Notation was tested in interviews. The $\bar{a}$ notation was used exclusively for the closed-form rounds and understood by all 30 interviewed students regardless of the notation used in their classes.
7. According to Baker's (2001) guidelines.
8. These data were collected as part of the NSF awarded project Teaching Inquiry-oriented Materials: Establishing Supports Award Number: 1431595; principal investigator: Estrella Johnson, Virginia Tech.

References

American Educational Research Association, American Psychological Association, & National Council on Measurement in Education. (2014). *Standards for educational and psychological testing*. Washington, DC: American Educational Research Association.

American Society for Biochemistry and Molecular Biology. (2015, February 2). *Concept inventories*. Retrieved from www.asbmb.org/uploadedFiles/Education/ TeachingStrategies/Concept_Inventory/Concept%20Inventories%202%20 2%202015.pdf

Andrews-Larson, C., Can, C., & Angstadt, A. (in press). *Proceedings of the 21st annual conference on research in undergraduate mathematics education*.

Baker, F. B. (2001). *The basics of item response theory*. College Park, MD: ERIC Clearinghouse on Assessment and Evaluation.

Blaich, C. F., & Wise, K. S. (2011). *The Wabash National Study: The impact of teaching practices and institutional conditions on student growth*. Paper presented at the annual meeting of the American Educational Research Association, New Orleans, LA.

Burn, B. (1996). What are the fundamental concepts of group theory? *Educational Studies in Mathematics, 31*(4), 371–377. doi:10.1007/BF00369154

Carlson, M., Oehrtman, M., & Engelke, N. (2010). The precalculus concept assessment: A tool for assessing students' reasoning abilities and understandings. *Cognition and Instruction, 28*(2), 113–145. doi:10.1080/07370001003676587

Cohen, J. (1992). A power primer. *Psychological Bulletin, 112*(1), 155.

Cronbach, L. J. (1988). Five perspectives on validity argument. *Test Validity*, 3–17.

Dalkey, N. C., & Helmer, O. (1963). An experimental application of the Delphi method to the use of experts. *Management Science, 9*(3), 458–467. doi:10.1287/ mnsc.9.3.458

Dubinsky, E., Dautermann, J., Leron, U., & Zazkis, R. (1994). On learning fundamental concepts of group theory. *Educational Studies in Mathematics*, 27(3), 267–305. doi:10.1007/BF01273732

Dubinsky, E., Dautermann, J., Leron, U., & Zazkis, R. (1997). A reaction to burn's "what are the fundamental concepts of group theory?" *Educational Studies in Mathematics*, 34(3), 249–253. doi:10.1007/BF01273732

Epstein, J. (2013). The calculus concept inventory-measurement of the effect of teaching methodology in mathematics. *Notices of the American Mathematical Society*, 60(8), 1018–1027. doi:10.1090/noti1033

Fraleigh, J. B. (2003). *A first course in abstract algebra* (7th ed.). Boston, MA: Pearson.

Gallian, J. A. (2013). *Contemporary abstract algebra* (8th ed.). Boston, MA: Brooks/Cole, Cengage Learning.

Gilbert, L., & Gilbert, J. (2009). *Elements of modern algebra* (7th ed.). Belmont, CA: Cengage Learning.

Gleason, J., White, D., Thomas, M., Bagley, S., & Rice, L. (2015). The calculus concept inventory: A psychometric analysis and framework for a new instrument. In T. Fukawa-Connelly, N. Infante, K. Keene, & M. Zandieh (Eds.), *Proceedings of the 18th annual conference on research in undergraduate mathematics education* (pp. 135–149). Pittsburgh, PA.

Hazzan, O. (1999). Reducing abstraction level when learning abstract algebra concepts. *Educational Studies in Mathematics*, 40(1), 71–90. doi:10.1023/a:1003780613628

Hazzan, O., & Leron, U. (1996). Students' use and misuse of mathematical theorems: The case of Lagrange's theorem. *For the Learning of Mathematics*, 16(1), 23–26.

Hestenes, D., Wells, M., & Swackhamer, G. (1992). Force concept inventory. *The Physics Teacher*, 30(3), 141–158. doi:10.1119/1.2343497

Hiebert, J., & Lefevre, P. (1986). Conceptual and procedural knowledge in mathematics: An introductory analysis. *Conceptual and Procedural Knowledge: The Case of Mathematics*, 2, 1–27.

Hungerford, T. W. (2013). *Abstract algebra: An introduction* (3rd ed.). Boston, MA: Brooks/Cole, Cengage Learning. Institute of Education Sciences, & National Science Foundation.

Kane, M. (2004). Certification testing as an illustration of argument-based validation. *Measurement*, 2(3), 135–170.

Lajoie, C., & Mura, R. (2000). What's in a name? A learning difficulty in connection with cyclic groups. *For the Learning of Mathematics*, 20(3), 29–33.

Lindell, R. S., Peak, E., & Foster, T. M. (2007, January). Are they all created equal? A comparison of different concept inventory development methodologies. In *AIP Conference Proceedings* (Vol. 883, No. 1, pp. 14–17). AIP. Retrieved from https://aip.scitation.org/doi/10.1063/1.2508680

Melhuish, K. M. (2015). *The design and validation of a group theory concept inventory* (Doctoral dissertation). Retrieved from http://pdxscholar.library.pdx.edu/open_access_etds/2490

Melhuish, K. M. (2018). Three conceptual replication studies in group theory. *Journal for Research in Mathematics Education*, 49(1), 9–38. doi:10.5951/jresematheduc.49.1.0009

Mesa, V. (2004). Characterizing practices associated with functions in middle school textbooks: An empirical approach. *Educational Studies in Mathematics, 56*(2–3), 255–286. doi:10.1023/b:educ.0000040409.63571.56

Messick, S. (1995). Validity of psychological assessment: Validation of inferences from persons' responses and performances as scientific inquiry into score meaning. *American Psychologist, 50*(9), 741. doi:10.1037//0003-066x.50.9.741

Novotná, J., & Hoch, M. (2008). How structure sense for algebraic expressions or equations is related to structure sense for abstract algebra. *Mathematics Education Research Journal, 20*(2), 93–104. doi:10.1007/bf03217479

Oktaç, A. (2016). Abstract Algebra Learning: Mental structures, definitions, examples, proofs and structure sense. *Annales de Didactique et de Sciences Cognitives, 21*, 297–316.

O'Shea, A., Breen, S., & Jaworski, B. (2016). The development of a function concept inventory. *International Journal of Research in Undergraduate Mathematics Education, 2*(3), 279–296. doi:10.1007/s40753-016-0030-5

Schilling, S. G., & Hill, H. C. (2007). Assessing measures of mathematical knowledge for teaching: A validity argument approach. *Measurement, 5*(2–3), 70–80. doi:10.1080/15366360701486965

Streveler, R. A., Olds, B. M., Miller, R. L., & Nelson, M. A. (2003, June). Using a Delphi study to identify the most difficult concepts for students to master in thermal and transport science. In *Proceedings of the annual conference of the American Society for Engineering Education.*

Tallman, M. A., Carlson, M. P., Bressoud, D. M., & Pearson, M. (2016). A characterization of calculus I final exams in US colleges and universities. *International Journal of Research in Undergraduate Mathematics Education, 2*(1), 105–133.

Travers, K. J., & Westbury, I. (1989). *The IEA study of mathematics I: Analysis of mathematics curricula.* Elmsford, NY: Pergamon Press.

Wasserman, N. H. (2016). Abstract algebra for algebra teaching: Influencing school mathematics instruction. *Canadian Journal of Science, Mathematics and Technology Education, 16*(1), 28–47. doi:10.1080/14926156.2015.1093200

Weber, K., & Alcock, L. (2004). Semantic and syntactic proof productions. *Educational Studies in Mathematics, 56*(2–3), 209–234. doi:10.1023/B:EDUC.0000040410.57253.a1

Weber, K., & Larsen, S. (2008). Teaching and learning group theory. *Making the Connection: Research and Teaching in Undergraduate Mathematics Education, 73*, 139. doi:10.5948/UPO9780883859759.012

Williams, B., Onsman, A., & Brown, T. (2010). Exploratory factor analysis: A five-step guide for novices. *Australasian Journal of Paramedicine, 8*(3).

Wladis, C., Offenholley, K., Licwinko, S., Dawes, D., & Lee, J. (2018). In A. Weinberg, C. Rasmussen, J. Rabin, M. Wawro, & S. Brown (Eds.), *Proceedings of the 21st Annual Conference on Research in Undergraduate Mathematics Education*, San Diego, California.

Zaslavsky, O., & Peled, I. (1996). Inhibiting factors in generating examples by mathematics teachers and student teachers: The case of binary operation. *Journal for Research in Mathematics Education*, 67–78. doi:10.2307/749198

7 Developing a Construct Map for Teacher Attentiveness

Michele B. Carney, Tatia Totorica, Laurie O. Cavey, and Patrick R. Lowenthal

Introduction

Construct maps, as described by Wilson (2005) in *Constructing Measures*, are used to describe an idea or concept that can be thought of as consisting of hierarchically ordered and qualitatively distinguishable categories along a continuum (e.g., low to high levels of knowledge for a particular topic). Construct maps are important tools in educational assessment and can serve multiple purposes related to development and validation as well as score interpretation and use. For example, construct maps are particularly useful for demonstrating how a theoretical construct has been operationalized into an assessment. Operationalizing a construct into an assessment necessarily bounds the construct into a smaller subset of ideas than is typically envisioned at the theoretical level. The construct map helps test developers and end users better understand what is and is not included within the assessment of a construct. Relatedly, Shepard (2018) has stated, "To support student learning, quantitative continua must also be represented substantively, 'describing in words and with examples what it looks like to improve in an area of learning' (National Research Council, 2001, p. 137)" (p. 1). Construct maps can be used to assist in the development of these substantive qualitative descriptions of test performance, which in turn provide meaningful score interpretation and use. Construct maps can also be helpful in the iterative improvement of an assessment, which might include informing continued item development efforts as testing needs change. These are just a few of the potential applications of construct maps for the improvement of assessments and their applications in education.

While Mark Wilson's work at the Bear Evaluation and Assessment Research center is extremely well-regarded, there are few examples within the measurement literature at-large specifying how to go about creating construct maps, particularly for open-ended assessment items. The purpose of this chapter is to provide an example of developing a construct map for teacher attentiveness to student thinking, referred to as *attentiveness*. Attentiveness is a particularly complex construct to assess, as it is related to the pedagogical content knowledge a teacher uses to respond to a student in ways that both take the student's ideas seriously and enable the student to build on their own ideas. Providing an example of this type of work for others demonstrates how complex constructs such as attentiveness can be mapped along a continuum via the use of responses to open-ended assessment items.

Literature Review

This section provides a review of the literature that informs the attentiveness construct, describes why it is important to assess attentiveness, and highlights key considerations in the assessment of attentiveness. The iterative relationship between construct map development, instrument development, and analysis and validation of an assessment tool for attentiveness is outlined.

The Attentiveness Construct

Attentiveness builds upon ideas from professional noticing (Jacobs, Lamb, & Philipp, 2010), formative assessment (Black & Wiliam, 2009), mathematical knowledge for teaching (Ball, Thames, & Phelps, 2008; Shulman, 1987), and progressive formalization (Freudenthal, 1973; Gravemeijer & van Galen, 2003; Treffers, 1987). It is defined as the *ability to analyze and respond to a particular student's mathematical ideas from a progressive formalization perspective.* It most closely parallels ideas from professional noticing—a set of interrelated skills for teaching that involve attending, interpreting, and responding to student ideas—but differs in that the focus with attentiveness is on *individual* student's thinking and ways in which student ideas can be built upon to become progressively formal. Attentiveness can be seen as a component of high-quality professional noticing but does not include many of the classroom-level components often attributed to professional noticing. Attentiveness can also be viewed as a significant contributor to formative assessment, as it helps bridge the processes of "engineering effective classroom discussions and other learning tasks that elicit evidence of student understanding" and "providing feedback that moves learners forward" within the seminal formative assessment framework developed by Black and Wiliam (Black & Wiliam, 2009). Bounding attentiveness to focus on an individual student's thinking allows the construct to be operationalized at a grain size appropriate for traditional forms of assessment (e.g., tests), whereas professional noticing and formative assessment, due to their classroom context, are harder to assess using traditional forms of assessment.

Attentiveness can be thought of as making use of components within the construct of mathematical knowledge for teaching (MKT), an oft-cited construct in the teacher education literature that refers to content and pedagogical knowledge specific to the work of teaching mathematics

(Ball et al., 2008; Shulman, 1987). MKT encompasses both traditional mathematical knowledge related to instructional content and knowledge that is specifically related to designing and managing students' classroom experiences. This includes knowledge of how students' mathematical ideas develop, how to promote the development of students' ideas, how to recognize common conceptualizations (both informal and formal), and how to identify and create mathematical tasks that elicit important mathematical ideas (Gravemeijer, Cobb, Bowers, & Whitenack, 2000; National Research Council, 2001; Stein, Engle, Smith, & Hughes, 2008; Stein, Smith, Henningsen, & Silver, 2000).

Attentiveness makes use of components within MKT because attending to students' thinking requires a teacher to (1) examine student work in relation to the mathematical intent of the task, (2) situate the student's work within a larger progression of student understanding for the topic, and (3) press students to generalize, formalize, or revise their ideas. In this way, attentiveness is bounded by employing a particular pedagogical lens (progressive formalization), which emphasizes the importance of building upon students' ideas (Freudenthal, 1973, 1991; Treffers, 1987). In progressive formalization, students initially apply their informal mathematical knowledge to a mathematically demanding task (Freudenthal, 1991). Students gradually develop more sophisticated ways of reasoning, eventually connecting their conceptual models to established conventions through the support of the teacher as they work through a sequence of mathematical tasks. From a progressive formalization perspective, a teacher's ability to analyze student work and respond in ways likely to support students in developing more sophisticated mathematical understanding is paramount to that teacher's practice.

Implementing progressive formalization requires the teacher to recognize the valid mathematical ideas within students' informal or incorrect answers and to make connections from students' informal ideas to more formal mathematics. Likewise, the teacher must be able to interpret incorrect answers that reveal a disconnect within the student's understanding and build on what the student *does* understand to help bridge to the targeted mathematical idea. In this way, attentiveness can be viewed in relation to high-leverage or "core" practices—regular routines that teachers use to engage students in productive work (Grossman, Hammerness, & McDonald, 2009). The *Learning to Teach In, From and Through Practice Project* (Lampert et al., 2013) has identified eliciting and responding to students' contributions as a significant practice that can be developed in teacher candidates. Attentiveness requires careful attention to students' ideas while keeping the important mathematical goals of the task in mind, similar to the type of ambitious teaching described by Lampert and her colleagues.

There have been multiple calls for mathematics pedagogy focused on these concepts. For example, *Principles to Actions* (National Council of

Teachers of Mathematics, 2014) calls for teachers to *elicit and use evidence of student thinking* as one of eight mathematics teaching practices; similarly, one of the high-leverage practices from TeachingWorks focuses on *working with individual students to elicit, probe, and develop their thinking about content* (TeachingWorks, 2018). Underlying these various calls is the need for teachers to be able to meaningfully interpret student work and build upon student thinking in ways that leads to their understanding of important mathematical ideas. While the goal is for teachers to interpret and respond to student thinking within the context of a classroom setting (with multiple competing priorities), attentiveness, with its narrowed focus on interpreting and responding to individual student work samples, is an appropriate starting place for teacher educators seeking to build teacher candidates' ability to engage in the wide range of important pedagogical practices focused on the progressive formalization of student thinking.

Assessing Attentiveness

Our instrument development work is, in part, a response to the need to hold teacher preparation and professional development programs accountable for the work they do in preparing and supporting teachers (e.g., Council for the Accreditation of Educator Preparation, 2015; Grossman et al., 2009; Lampert, 2009). High-quality instruments, designed for research and program evaluation, are essential to this endeavor. While commercial assessments like the PRAXIS have long been used for individual evaluation of content knowledge, there has been a growing interest in developing instruments of constructs central to teaching mathematics (e.g., Learning Mathematics for Teaching, 2005). Such instruments could enable educators to examine changes over time for a single program or to make comparisons across multiple programs. For teacher educators whose work focuses on developing mathematics teachers' ability to analyze and respond to student thinking, tools are needed to help identify whether teachers' engagement in course and program activities influences their attentiveness.

An important consideration when constructing instruments to measure a complex construct such as attentiveness is the meaningful operationalization of the construct. Performance assessments enacted in practice provide an authentic means of assessing constructs such as attentiveness; however, they are difficult to implement at scale and costly in terms of time and money to score accurately. More traditional forms of assessment provide a more efficient means of assessing attentiveness but can be compromised in terms of authenticity and connections to practice. Given both a program-level focus and a goal of teachers effectively analyzing and responding to student thinking in practice, a scalable means of assessing attentiveness is needed. Therefore, we opted to develop a more

traditional form of assessment, with a focus on authenticity in the operationalization of attentiveness.

Our operationalization of attentiveness in the Attentiveness in Quantitative Reasoning Inventory (QRI) involves identification of key disciplinary ideas for the mathematics topic, in this case quantitative reasoning,
in conjunction with common, informal ways students reason about these
topics. The key disciplinary ideas and ways of student reasoning are
examined through the lens of the following professional noticing categories: analyzing the approach, interpreting understanding, and responding
to the student. While we use the professional noticing categories, our
operationalization differs slightly from professional noticing in that our
focus is on teacher candidates' ability to be attentive to an individual
student's thinking, which we see as a precursor to the broader perspective
of professional noticing in a classroom context. In addition, our analysis of the assessment item responses includes a focus on analyzing the
approach, interpreting understanding, and responding to the student in
ways that builds upon their thinking and allows for a progressive formalization of their understanding.

The current version of the Attentiveness in QRI makes use of a
constructed-response item format. This version of the instrument is useful
for program level assessment but can be time-consuming to analyze and
score if administered on a large scale. This limitation is further addressed
in the "Discussion" section under "Further Assessment Development."

Construct Map Development

In Wilson's (2005) book, *Constructing Measures*, creating a construct
map is the recommended first step in assessment development. An important decision involved in this first step is determining whether to first
articulate the qualitative ordering of the responses or that of the respondents. Wilson states:

> In creating a construct map, the measurer must be clear about whether
> the construct is defined in terms of who is to be measured—the
> respondents—or what responses they might give—the item responses.
> Eventually both will be needed, but often it makes sense in a specific
> context to start with one rather than the other.
>
> (p. 38)

In the development of the instrument described in this chapter, we chose
to focus on the qualitative ordering of item responses. While we do not
describe the response space for respondents in detail here, it is an important aspect of supporting score interpretation and use that is addressed at
the end of the chapter.

In the next section, we address our methods and results associated with
articulating the item response portion of the construct map for teacher

attentiveness, with a focus on providing detailed examples of how to develop the qualitative categories and ordering of the responses. The first step was a review of the literature to identify potential shifts in responses as attentiveness increases. The next steps were to code and categorize the responses received from teachers and teacher candidates to further delineate and hierarchically organize the categories of attentiveness. Finally, this information was organized into a construct map diagram for attentiveness.

Methods and Results

Context

The National Science Foundation funded the Video Case Analysis of Student Thinking (VCAST) project. The purpose of the VCAST project is to develop instructional materials with the potential to increase secondary mathematics teacher candidates' ability to analyze and respond to student thinking in quantitative reasoning contexts. In an effort to evaluate the influence of the VCAST intervention on teacher candidates' attentiveness, the Attentiveness in QRI was administered in the fall of 2017 to candidates enrolled in an upper-division mathematics course addressing functions and modeling at the secondary level. The Attentiveness in QRI was also administered in the summer of 2018 to assist in the evaluation of a professional development institute that utilized the VCAST materials and focused on functions and modeling for secondary teachers.

Participants

The Attentiveness in QRI was administered to a total of 42 respondents: 17 secondary mathematics teacher candidates and 25 secondary mathematics teachers. The teacher candidates were enrolled in an upper-division functions and modeling mathematics course designed for future secondary mathematics teachers in the United States. The mathematics teachers teach grade levels ranging from middle school to high school in the United States and were enrolled in a state-funded professional development course offered over the summer. The Attentiveness in QRI responses were collected prior to and following an intervention designed to improve teacher candidates' and teachers' attentiveness. The timing of the instrument administration—prior to and following intervention—was intended to elicit a broad range of responses along the attentiveness continuum.

Instrument

The Attentiveness in QRI is being developed using the multiphase process and general item framework described by Carney, Cavey, and Hughes (2017). See the Appendix for a description of the development process and purpose statement for the instrument. The instrument used for data collection for this study involved collecting constructed-responses (typically short paragraphs) to item prompts. The instrument includes three quantitative reasoning tasks with two student work samples per task along with prompts related to (a) the mathematical intent of the task, (b) describing the student approach, (c) describing the student understanding, and (d) describing how the test taker would respond to the student based on the student work sample. The prompts vary slightly across items depending on the task and student work sample.

More specifically, in terms of the instrument, clarification of the differences between the assessment item, the task, the student work sample, and prompts is provided. The term *assessment item* refers to a student task, its related student work samples, and item prompts. There are three total assessment items in the Phase 3 Attentiveness in QRI. *Task* refers to the quantitative reasoning problem presented to secondary (grades 6–12) mathematics students. There are three tasks (one for each assessment item) embedded in the Attentiveness in QRI, each with a different mathematical focus: (a) estimating a line of best fit, (b) proportional reasoning, and (c) estimating total distance traveled from a time and speed graph. The tasks were designed to be accessible across ability levels and to elicit a broad range of student approaches. In other words, they were designed so those without formal mathematical knowledge could still work productively toward correct solutions. *Student work sample* refers to the solution process (written and/or verbal) generated by the secondary mathematics student. Two authentic student work samples are presented with their accompanying prompts. The student work samples feature elements that can be considered correct, elements that can be considered incorrect, and/or relatively informal approaches to solving the task. This intentional selection of these types of student work samples and the tasks which elicit them is informed by a pedagogical lens of progressive formalization and the need for teachers to be able to understand and build upon students' informal understandings. *Prompts* refers to the wording used within the instrument to generate a response from the test taker. The prompts in the Attentiveness in QRI vary across assessment items. For instance, the prompts associated with estimating a line of best fit and estimating total distance traveled tasks primarily focus on the following four areas:

- The important mathematical idea(s) the task is targeting (describing intent);
- Description of the approach used by the student (describing student approach);
- Description of what the student response reveals about their understanding of the important mathematical idea(s) in this task (describing student understanding);
- Description of how the test taker would respond to the student (teacher response).

The proportional reasoning task, on the other hand, only has two prompts focused on explaining the similarities and differences in understanding between the two student work samples.

Figure 7.1 presents the item related to estimating total distance traveled from a time and speed graph. This assessment item represents the typical format of our attentiveness items and is very similar in structure to the line of best fit task.

Data

In this sample, responses to the assessment prompts were typically one to three sentences in length. Certain prompts, such as describing the mathematical intent of the task, lend themselves to shorter phrases, while other prompts, such as describing how the test taker would respond to the student, lend themselves to short paragraph answers.

A total of 39 respondents had two full sets of responses; one set of responses was collected prior to participation in an attentiveness intervention and the other set of responses was collected following such

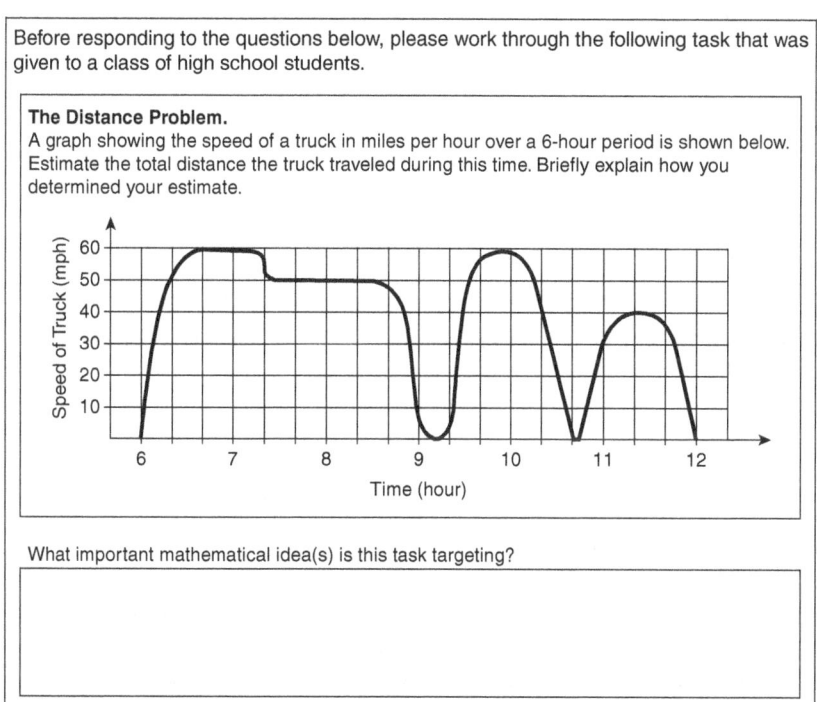

Figure 7.1 Example assessment item consisting of the interpreting a time and speed graph to estimate distance traveled task, student work samples, and related prompts

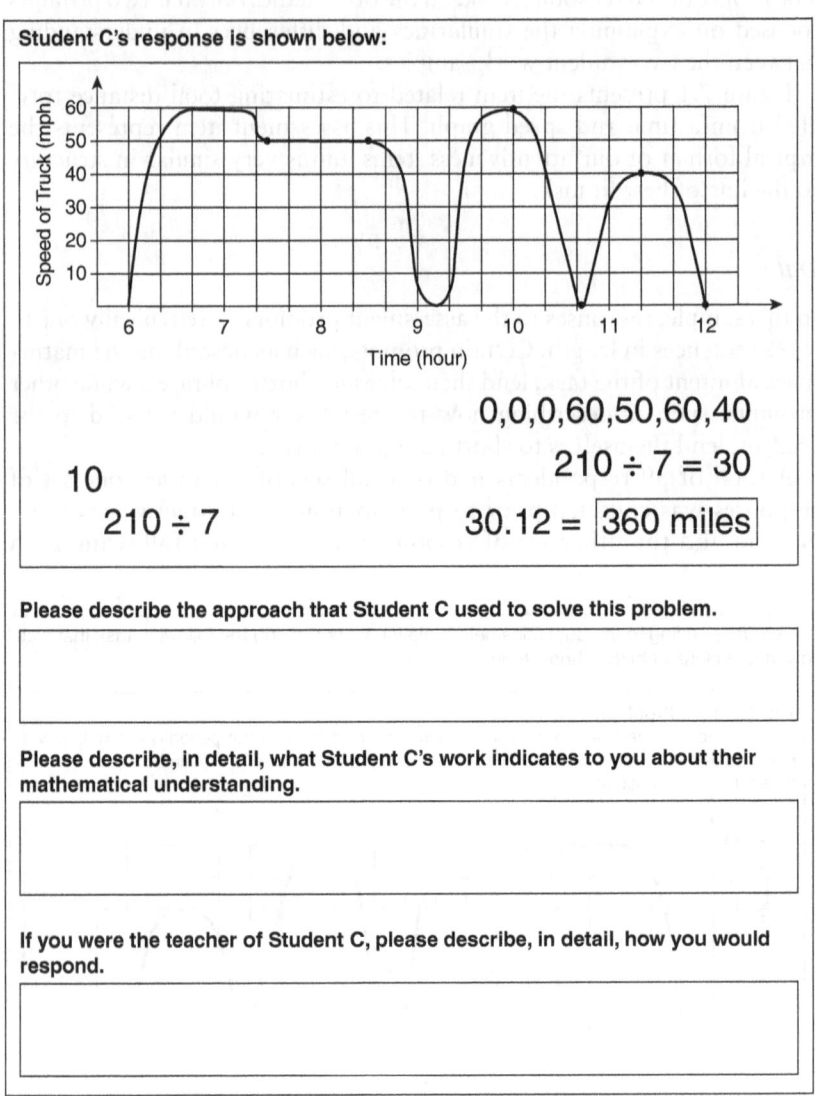

Student C's response is shown below:

0,0,0,60,50,60,40

10

210 ÷ 7

210 ÷ 7 = 30

30.12 = 360 miles

Please describe the approach that Student C used to solve this problem.

Please describe, in detail, what Student C's work indicates to you about their mathematical understanding.

If you were the teacher of Student C, please describe, in detail, how you would respond.

Figure 7.1 (Continued)

participation. Three additional sets of responses were collected prior to the attentiveness intervention with teachers. This resulted in 81 sets of responses for analysis. Each set consisted of responses to 15 total prompts across the three assessment items for a total of 1,215 responses to prompts. Our unit of analysis is at the response-to-prompt level.

Prior to analysis, data were de-identified for all respondent information, including respondent name and timing of assessment responses (i.e., pre- vs. post-attentiveness intervention). A randomly generated ID

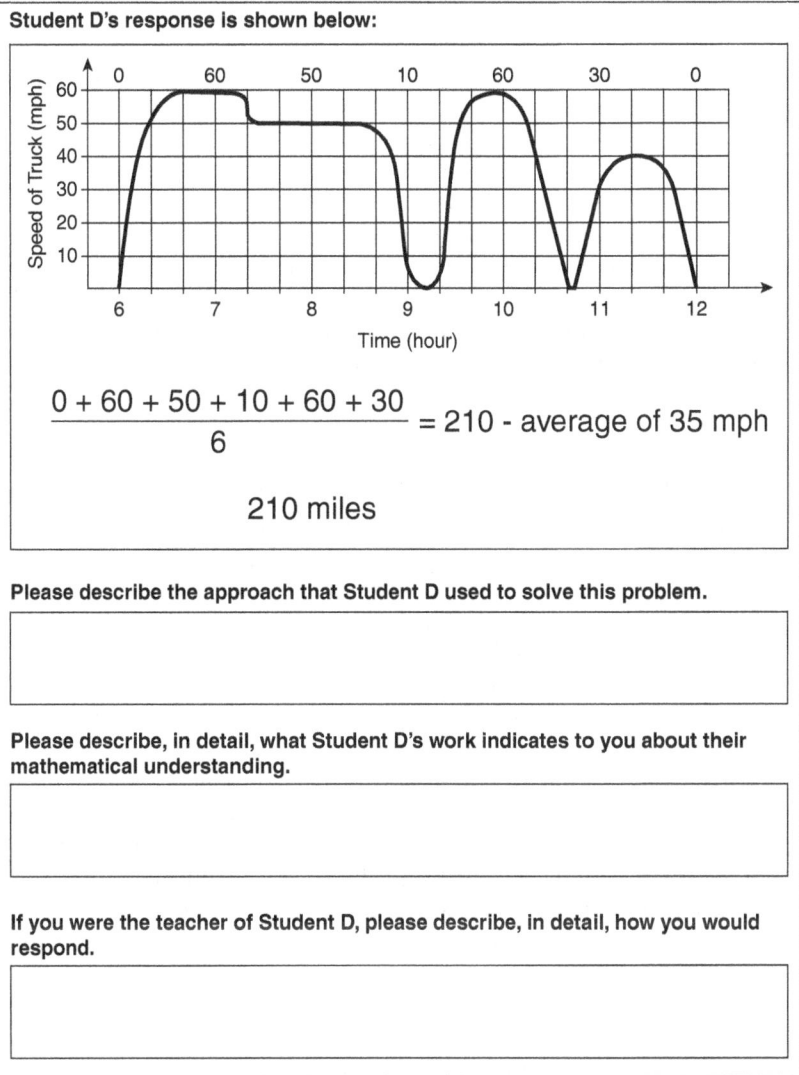

Student D's response is shown below:

$$\frac{0 + 60 + 50 + 10 + 60 + 30}{6} = 210 \text{ - average of 35 mph}$$

210 miles

Please describe the approach that Student D used to solve this problem.

Please describe, in detail, what Student D's work indicates to you about their mathematical understanding.

If you were the teacher of Student D, please describe, in detail, how you would respond.

Figure 7.1 (Continued)

number was associated with each set of responses so the information could be linked back to respondents following coding.

Construct Map Development Process

The construct development process for teacher attentiveness occurred sequentially in four steps: (1) initial construct map and Attentiveness in QRI development based on the literature review, (2) structural coding of

Attentiveness in QRI response data, (3) code mapping to categories, and (4) scoring and construct map diagram. Because completion of each step was a precursor to the next, analysis and results are reported for each step individually rather than for the process as a whole. We have organized the information in this section around the four steps.

Literature Analysis

Prior to analysis of the data, the research team examined existing coding schemes, primarily from the professional noticing literature, to develop an initial conceptualization for attentiveness unrelated to a specific mathematical topic. The primary focus of the literature review was to identify potential shifts between levels of attentiveness. These findings are summarized below:

- Teacher shifts from providing non-specific descriptions of a student approach to (a) detailing what is mathematically correct and incorrect about a solution strategy, and/or (b) differentiating between the aspects of the student response that are generalizable and non-generalizable (Bartell, Webel, Bowen, & Dyson, 2013; Jacobs et al., 2010; Sherin & van Es, 2005; Stockero, 2008);
- Teacher shifts from under- or overgeneralizing student's understandings *to* describing the student's understanding related to both the particular context's quantitative reasoning demands and to generalizations beyond the context (Bartell et al., 2013; Jacobs et al., 2010);
- Teacher shifts from accepting or affirming student responses, pressing toward an answer without probing thinking, or asking generic questions *to* (a) posing questions or prompts that make use of the student's reasoning to further probe or help students clarify their thinking, and/or (b) offering suggestions for next problems that press particular aspects of quantitative relationships (Jacobs et al., 2010; van den Kieboom, Magiera, & Moyer, 2014).

Using the research literature to identify shifts toward increasing levels of attentiveness to student thinking provided some initial concepts for consideration and supported an assessment design which targeted three of the four primary components of attentiveness (describing student approach, describing student understanding, and teacher response). The fourth component, which represents a teacher's ability to identify and articulate the important mathematical idea(s) a task is targeting (describing intent), affords differentiation between those whose analysis of student work is appropriately aligned to the intent of the task and the grade level for which the task is used. The next phase of the construct map development, structural coding, focuses on analysis of test takers' qualitative responses to the Attentiveness in QRI prompts. For illustration purposes and to accommodate space constraints, we focus on the analysis of test taker responses elicited by assessment item prompts which target the attentiveness component of *describing student understanding*.

Structural Coding

Structural coding identifies and applies conceptual phrases to segments of data typically collected via interview or constructed response (Saldaña, 2015). This step of analysis involved an inductive, open-coding approach during which the researchers identified emergent themes for attentiveness in test takers' responses specific to the instrument prompt. Two researchers worked in multiple iterative rounds of independent and collaborative coding to develop, apply, and revise the coding schemes.

Prior to examining any data, the researchers brainstormed features of the desired exemplar responses each prompt was intended to elicit and a potential range of responses that the shifts identified above might forecast. This served to focus their lens on potential indicators of attentiveness before analysis began.

In round one of data analysis, the researchers independently read through a subset of the teacher responses and used open-coding methods to identify emergent themes. They then met to discuss noticed themes and to reach consensus on an initial coding scheme.

For round two, the researchers again independently coded, taking note of when challenges with the initial coding scheme arose. When roughly a third of these data had been independently coded, the researchers met to check for inter-coder agreement and revise the coding scheme as necessary. Responses with different codes were discussed among coders until agreement was reached on the meaning of the code and how that code was evident in the response.

This iterative process of identifying, independently applying, and collaboratively revising a coding scheme for a particular prompt often entailed three to four rounds before full consensus was met across all responses. During this process, coding tables were developed to describe and summarize the codes at the individual prompt level (15 prompts, therefore 15 tables). These tables include the code name, code description, and one to three exemplar responses.

The example presented comes from responses to the "Describe Student C's understanding" prompt for the estimating total distance traveled assessment item (see Figure 7.1). Student C's work sample provided multiple aspects of understanding, both correct and incorrect, to which test takers could attend. For example, Student C appears to understand that calculating the total distance involves multiplying the amount of time traveled by an estimate for the average speed during that time. Responses that addressed this idea were labeled with the code *Distance Calculation*. Student C's selection of speeds (*Points*) to average does not include explicit consideration of the amount of time for which that speed was traveled. In addition, the student's calculation of the amount of time traveled is inaccurate. Responses that addressed these ideas were labeled with the codes *Duration of Speed* and *Total Time*, respectively. Table 7.1 presents an overview of the full set of code names, descriptions, and selected exemplars related to the Student C understanding prompt. This example is typical of the coding tables that were developed across the 15 prompts.

In addition to the identification of codes, responses to this prompt also indicated there were respondents who focused on (a) describing what Student C did understand, (b) describing what Student C did not understand, and (c) describing both what Student C did and did not understand. Recognizing this pattern was useful in the next step, which is code mapping to categories.

Code Mapping to Categories

The next phase of analysis involved code mapping (Saldaña, 2015) across the collective code tables generated for common prompt types (four prompt types: describing intent, describing student approach, describing student understanding, and teacher response). This involved identifying and naming categories and subcategories across common prompt types that would encompass the prompt-specific codes in each individual table. Again, the desired shifts in attentiveness described by the literature helped to inform this work. The example provided here describes the development of the hierarchical categories for the five coding tables related to the *describing student understanding* prompts.

These coding tables for the describing student understanding prompts were examined to identify categories and subcategories that consistently threaded across the codes, regardless of the mathematical task and student work samples. Two primary categories were identified: *mathematical focus* and *perspective on understanding*. The mathematical focus category primarily related to the specificity of the claim made about student understanding in relation to the mathematical intent of the task. There were four subcategories; supported specific claim, supported non-specific claim, unsupported claim, and no claim. The perspective on understanding category related to whether the claim about student understanding focused on what the student understood and/or did not understand. The two subcategories were affordances *and* constraints, which featured a dual perspective and involved inference language related both to what the student did and did not understand, while the subcategory affordances *or* constraints involved a single perspective and included inference language related either to what the student did or did not understand.

The coding tables for the describing student understanding prompt were used to inform the development of the categories and subcategories related to the student understanding component of attentiveness.

Table 7.1 Coding Table for the Student C Understanding Prompt for the Estimating Total Distance Traveled Assessment Item

Estimating total distance traveled from a time and speed graph: Describing student understanding

Code Name	Code Description—Candidate response to	Code Exemplar(s)
Describes aspects of what the student does understand		
Points	Indicates that the student point selection is likely not random but they don't necessarily articulate the likely reasoning.	—I am unsure of how they chose there [*sic*] 7 points to look at though.
Distance Calculation	Indicates the student knows or understands a velocity (or speed) should be multiplied by a time value to get distance.	—They understand that to find distance they need to multiply a time by a speed. —They have the basic idea right of multiplying time and velocity to get distance.
Average	References student understanding of average (related to calculation involving the 7 points).	—They understand how to average things in this sense of add them up and then divide by how many there are.
Overgeneralize	Overgeneralizes student understanding in ways that are not supported by the evidence.	—Student C understands how the axes and their units relate to each other, this is shown by multiplying the mph by the hours to get miles.
Generic	Fails to explicitly reference either the student approach or connect to the task intent.	—It actually says a lot, this is a very interesting approach to this problem. —How to read a graph.
Describes aspects of what the student does NOT understand		
Duration of Speed	References student not taking into account the length of time a particular speed was traveled.	—They, however, missed some critical ideas about duration the vehicle was at certain speeds. —First and foremost, they do not account for the "unevenness" of their sections, so averaging them does not give a very good idea of the true average speed.

(Continued)

Table 7.1 (Continued)

Estimating total distance traveled from a time and speed graph: Describing student understanding

Code Name	Code Description—Candidate response to	Code Exemplar(s)
Total Time	References student mistake of using 12 (the last time marked on the x-axis) instead of 6 (the number of hours traveled) as the number of hours traveled.	—They misread the graph, as it starts at 6, so there are 6 total hours, not 12.
Alternative Approach	References an alternative approach the student could have used to solve the problem.	—They used points that represented change in speed rather than average speeds.
Overgeneralize	Overgeneralizes student (lack of) understanding in ways that are not supported by the evidence.	—I think Student C was in a hurry to get this problem done because they didn't pay much attention to the starting point on the x-axis and made some simple mistakes. It also would appear that they do not understand what they are doing.
Formal	Identifies potential formal mathematical (mis) understandings related to the task, but the language used (e.g., area under the curve) is beyond the scope of the task.	—Student C does not understand the relationship between velocity and distance, or at least does not connect the ideas of area under a velocity curve and distance traveled.
Disparate	References (mis)understandings involving related ideas (acceleration, position, direction of movement, etc.) that are not the focus of the task or explicitly evidenced in the student's work.	*—no exemplars*
Incorrect Mathematics	Indicates an incorrect understanding of mathematics, whether in the task itself or in the student's work.	—This indicates that they don't know how to interpret values based on the lines/curves of the graph.

However, when assigning a subcategory to a response, we found that additional considerations were necessary to account for the mathematical intent of each task. For example, the code *duration of speed* was subcategorized as a supported specific claim because the student process for approximating the average speed—including the amount of time a particular speed was traveled—was not only supported, but it was considered critical to the intent of the task and a key aspect of the understanding demonstrated by Student C's work sample. Similarly, the code *total time* tended to be subcategorized as supported non-specific because while it was evident the student did not account for the x-axis starting at 6 instead of 0, this was considered a less critical understanding in terms of the task's mathematical intent. Additional assignments of the specific versus non-specific subcategory required a more nuanced examination and discussion of the task's mathematical intent among coders, as a particular code from one coding table could be classified as specific while, were the student task for the assessment item to be changed, the same code could be classified as non-specific.

Table 7.2 presents the categories and subcategories identified for the describing student understanding prompt and includes exemplar responses to the assessment item involving the estimating total distance traveled task (see Figure 7.1) for Student C. This example is typical of the category tables developed across the four prompt types and supports the work for the final step of construct map development, which is scoring and construct map diagram.

Scoring and Construct Map Diagram

The next step in the process involved assigning numerical scores to the common categories and subcategories given to each response. Assigning numerical scores to categorical labels has both benefits and drawbacks. One serious drawback is that when a numerical score is assigned to qualitative data, a significant amount of information is lost in terms of the depth and detail of the response. However, a benefit of assigning a numerical score to the category labels is the ability to see more general trends in respondents' level of attentiveness. Sometimes the level of detail provided by coding and categorizing data can make it difficult to generate a more holistic picture of teachers' attentiveness. By attending to both aspects in our construct map development process, our qualitative interpretations of the overall quantitative score present a more nuanced perspective of teachers' attentiveness.

Following the development of the common categories and subcategories, research project personnel discussed the process of converting these to numerical scores. Within the mathematical focus category, responses categorized as supported specific claim were assigned 2 points, while responses categorized as supported non-specific claim were assigned 1

Table 7.2 Common Categories and Subcategories for Describing Student Understanding Prompts Including Exemplars From the Estimating Total Distance Traveled From a Time and Speed Graph for Student C

Category Name	Subcategory Name	Subcategory Description	Exemplars
Mathematical Focus	Supported specific claim	Makes an accurate inference about student understanding (or lack of understanding) supported by evidence that directly addresses the mathematical intent of the task and impacts student's ability to productively engage with the task.	They seem to understand that the question needs to be answered given the average speed of the truck over an interval of time, however they seem to be missing information about the increasing/decreasing speeds of the truck over different time intervals that would contribute to the overall distance traveled.
	Supported non-specific claim	Makes an accurate inference about student understanding (or lack of understanding) that is supported by evidence but does not address the mathematical intent of the task.	They understand how to find the average of a set of data.
	Unsupported claim	Makes an inference about student conceptual understanding (either what is understood or not understood) that is not supported by the evidence from student work. This includes overly formal and incorrect mathematics.	They do not understand that the area under the curve of velocity is the position.
	No claim	Describes the student's mathematics without making an inference about conceptual understanding or knowledge.	No exemplars for Student C
Perspective on Understanding	Affordances AND Constraints	The teacher or candidate uses inference language about what the student understands AND does not understand.	Student C understands how the axes and their units relate to each other, this is shown by multiplying the mph by the hours to get miles. Though they also read the x-axis wrong by assuming it started at zero then just looked at the last point of the graph to get 12 hours.
	Affordances OR Constraints	The candidate uses inference language about what the student understands OR does not understand.	They didn't think about whether if traveling at different speeds for different amounts of time would matter (constraint perspective).

point. Responses categorized as unsupported or no-claim were assigned 0 points. The point levels assigned to these subcategories were based both on our literature review and their likely impact on classroom instruction. For example, an unsupported claim made by a teacher could result in the teacher misdirecting the student, or at the very least, offering a suggested next step that would potentially be confusing for the student. Responses categorized as either an unsupported claim or as a no-claim were determined to be equally less beneficial and potentially an obstruction to supporting student understanding. Within the perspective on understanding category, responses categorized as affordances and constraints were assigned 1 point. Responses categorized as affordance or constraints were assigned 0 points. Similar to the mathematical focus category, the point levels assigned to the perspective on understanding subcategories were primarily based on their likely impact on classroom instruction. In this case, responses categorized with both affordance and constraint were determined to be potentially more beneficial to a student's understanding than responses categorized as either an affordance or constraint. Teachers who recognized both the productive and unproductive ideas in student work samples were seen as likely to build upon the student thinking in a way that would also target the mathematical intent of the task, while teachers who only described either the affordance or constraints were seen as likely to praise student work or correct the student misunderstanding, respectively.

Figure 7.2 presents the construct map for the describing student understanding component of attentiveness, based on the identified categories and subcategories, and includes the point assignments to each subcategory. The construct map diagram presented differs from Wilson's (2005) version of a construct map, as typically depicted in a table similar to Table 7.2, but the two construct maps are similar in that they present increasingly sophisticated ways of reasoning within a construct and have the potential to assist with meaningful score interpretation.

The construct map diagram in Figure 7.2 illustrates hypothesized relationships and, in conjunction with the category and subcategory descriptions from Table 7.2, provides the describing student understanding construct map for teacher attentiveness. It depicts the categories of mathematical focus and perspective on understanding as separate components of teachers' attentiveness to student understanding. However, it is important to note that we do not anticipate they will function as separate dimensions in our measurement model because they are so intertwined, and the nature of the relationships between these categories needs to be further examined. We represent them separately in the construct map diagram because the categories and subcategories are worth capturing for score interpretation and use in instructional applications. It is also worth noting that subcategories within a category are not necessarily aligned with the subcategory in an adjacent category. Instead, the subcategories

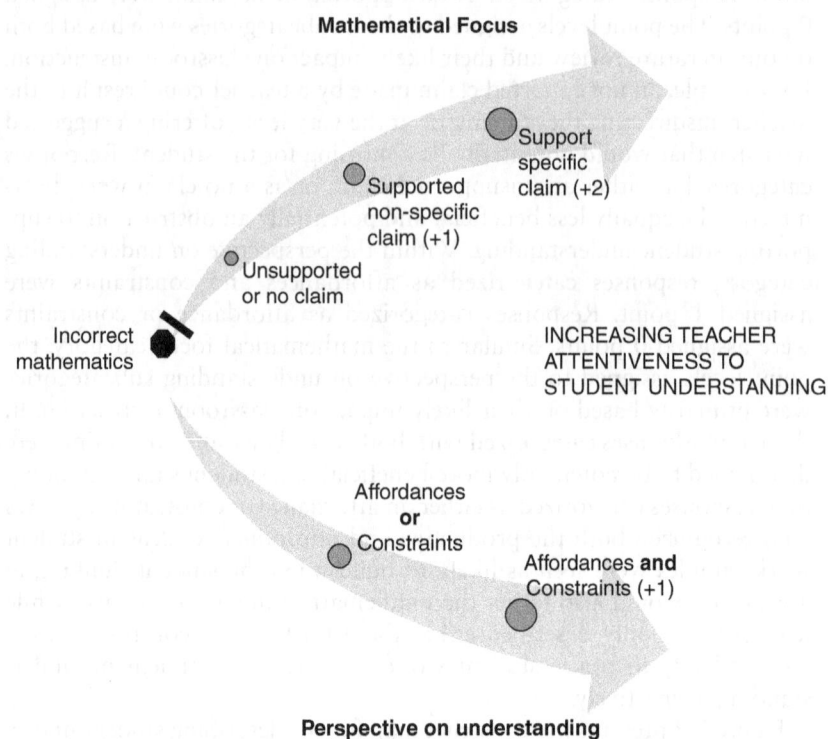

Figure 7.2 Describing student understanding construct map diagram for teacher attentiveness

are aligned with the arrow representing increasing/decreasing attentiveness. For example, some responses were categorized as affordances *and* constraints, the highest subcategory in the perspective on understanding category, while also being categorized as supported non-specific in terms of the mathematical focus. Other responses, due to the test taker's use of incorrect mathematics, did not score on the mathematical focus component at all, though they still provided evidence of a perspective on understanding. In terms of a teacher's ability to respond productively to students' thinking, a perspective that considers both what students understand and do not understand could potentially lead to dialogue with students where the teacher's incorrect mathematics is rectified.

Discussion

Developing a construct map for attentiveness serves multiple purposes and has numerous applications in terms of assessment development, validation, and score interpretations and uses related to instruction. We address three major purposes and their applications: score interpretation, construct representation, and further assessment development.

Score Interpretation

The construct map development process described in this chapter can be viewed as a response to recent calls for assessment developers to provide for substantive qualitative descriptions in relation to quantitative score continuum (Shepard, 2018), as the development of construct maps is an important first step in this work. The identified categories and subcategories for item responses and their relationships can be used to generate initial statements related to respondents' qualitatively different levels of attentiveness. For example, we anticipate scores in the upper third of the scoring range will be associated with respondents who consistently attend to the mathematics concepts evidenced in student work that involves articulation both of what the student understands and does not understand. This is based on scores in the upper third of the scoring range primarily being associated with responses from the subcategories of supported specific claim and affordances and constraints. Similarly, we anticipate scores from the middle third of the scoring range will be associated with respondents who explicitly attend to mathematics concepts evidenced in student work (subcategory supported specific claim) but only articulate what the student understands or does not understand (subcategory affordances or constraints), or they may have less specific attention to the mathematical intent of the task (subcategory supported non-specific) but focus on articulating what the student understands and does not understand (subcategory affordances or constraints). By articulating the likely combinations of response subcategories and point options illustrated by the construct map diagram, qualitative descriptors of respondents' scores are readily generated. These descriptors can then be further examined in relation to empirical data and refined to develop the qualitative ordering of respondents within the construct map. This work supports valid inferences regarding teachers' level of attentiveness based on their quantitative scores. Meaningful score interpretations and valid inferences also contribute to research being done in both formative assessment and professional noticing, as examining the relationships between attentiveness scores and the quality of noticing and assessment during mathematics instruction could reveal unexplored connections worthy of further study.

Construct Representation

The attentiveness construct draws from multiple aspects of the research literature, such as mathematical knowledge for teaching, progressive formalization, and professional noticing. The generation of construct maps that summarize and generalize the range of reasoning captured by our constructed-response items has two important applications. First, it helps to describe what is being assessed in our operationalized version of the construct, assisting with construction representation validation processes. As previously described, operationalizing a construct necessarily restricts the boundaries of that construct beyond what might be included in a theoretical description. Our construct map describes which aspects are and are not included within the operationalization. For example, the *describing student understanding* construct map clarifies the aspects of understanding teachers focus on in their response: mathematical focus and perspective on understanding. Providing a summary description of this work to assessment users can help them better understand the construct focus of the assessment to determine if it is appropriate for their needs. A second potential application involves the use of the construct map by mathematics teacher educators in instructional or professional development settings. Because the construct map illustrates and describes the components of attentiveness, it has the potential to support teacher learning and understanding of the construct itself and to frame and inform activities designed to support growth and development of attentiveness in educators.

Further Assessment Development

Lastly, regarding further assessment development, the process of categorizing and scoring constructed-responses can be time-consuming and expensive when applied to a large data set. One potential solution is further item development through the construction of selected-response items (for an example of this application, see Carney et al., 2017). The coding and category tables describe the range of reasoning that occurs in response to a particular assessment task, student work sample, and related prompt. Describing the range of reasoning elicited by constructed-response items makes it possible—as assessment developers—to identify authentic responses for use in the development of selected-response items that represent this range of reasoning found in the constructed-response version. This change in item type provides for a more efficient scoring process. By incorporating authentic responses generated by teachers or teacher candidates, we can be more confident that our selected-response items have the potential to (a) generate a full range of scores along the attentiveness continuum, (b) validly measure attentiveness in meaningful ways, and (c) inform future refinement of our construct operationalization. The codes and construct maps will also provide a point of comparison for data from cognitive interviews on the response process for the selected-response items to assist in determining how the change in item format potentially influences and/or further bounds the construct.

The purposes identified here are just a few examples of the usefulness of developing initial construct maps for applications related to assessment development, scoring, and validation, in addition to further developing our understanding of the operationalized version of the construct.

Limitations

It is important to realize that any time a construct is measured, the potential reasoning space encompassed by the theoretical construct becomes limited by the operationalization. The tasks, student work samples, and prompts all serve to provide a frame within which teachers can demonstrate their level of attentiveness. But we must realize these selections also limit the ways in which this demonstration occurs. Therefore, our construct map for attentiveness to student understanding is limited to the assessment from which it was created. It is possible that additional categories or subcategories of responses would emerge, had we operationalized the construct in a different manner.

Current and Future Work

In this section, we briefly describe our current and future work related to the Attentiveness in QRI and the ways in which the development of a construct map for attentiveness will inform other work related to the VCAST project, thereby illustrating some of our points in the discussion.

The construct map for attentiveness was recently used to inform modifications to the Attentiveness in QRI. In particular, the process of developing the construct map resulted in greater awareness of our intent for test takers to consciously attend to the context of item prompts. Explicit articulation of this intent occurred while code mapping to categories, as the code-mapping process revealed key similarities in the quality of responses across common prompt types. That is, for all four prompt types (describing intent, describing student approach, describing student understanding, teacher response), we realized we consistently assigned higher scores to responses that were informed by the mathematical level at which the task was used. This insight resulted in revising item prompts in an attempt to make our intention clearer to test takers. For example, the prompt "What mathematical ideas is this task targeting?" would be adjusted to "What Algebra I mathematical ideas is this task targeting?" for a constructed-response prompt (describe intent) when a task designed for Algebra I students is used.

In addition, the revised version of the Attentiveness in QRI now comprises selected-response items and addresses a broader range of quantitative reasoning tasks. The process of developing the construct map contributed to our ability to construct authentic responses. The exemplars included in our code tables were particularly helpful in the identification of responses, as they allowed us to confidently choose appropriate level responses for prompts. To further develop the Attentiveness in QRI, we plan to use Rasch modeling to study how the items are functioning psychometrically. Beyond testing for item functionality, we plan to draft substantive qualitative descriptions along the quantitative score continuum for attentiveness.

Finally, we intend to apply the construct map to assess the attentiveness intervention developed for the VCAST project. As part of the intervention, teacher candidates work through an online learning module which includes prompts directly related to operationalized components of the attentiveness construct (e.g., describing student understanding). Applying the construct map in this way will potentially lead to our ability to fine-tune our operationalization of attentiveness and thereby influence both the attentiveness instructional interventions and the Attentiveness in QRI.

Conclusion

The goal of this chapter was to illustrate how our instrument development process aimed at assessing teachers' attentiveness to secondary students' quantitative reasoning might provide insight to other researchers as they engage in their own assessment development efforts. This chapter builds upon Wilson's (2005) explication of construct maps by providing an example embedded within the assessment of a particularly complex construct. The focus includes, but also goes beyond, how construct maps can be used for assessment development. Generating construct maps and their associated category/code descriptors through analysis of authentic qualitative responses to constructed-response items has the potential to provide clarity on how the resultant operationalization bounds the construct, further develop understanding of theory, and provide a basis for the development of substantive qualitative interpretations of assessment scores along a quantitative continuum.

Author's Note

This research was supported in part by grant #1726543, *Preparing Secondary Mathematics Teachers With Video Cases of Students' Functional Reasoning* from the National Science Foundation.

References

Ball, D. L., Thames, M. H., & Phelps, G. C. (2008). Content knowledge for teaching: What makes it special? *Journal of Teacher Education, 59*(5), 389–407.

Bartell, T. G., Webel, C., Bowen, B., & Dyson, N. (2013). Prospective teacher learning: Recognizing evidence of conceptual understanding. *Journal of Mathematics Teacher Education, 16*(1), 57–79.

Black, P., & Wiliam, D. (2009). Developing the theory of formative assessment. *Educational Assessment, Evaluation and Accountability, 21*(1), 5–31.

Carney, M., Cavey, L., & Hughes, G. (2017). Assessing teacher attentiveness: Validity claims and evidence. *Elementary School Journal, 118*(2), 281–309.

Council for the Accreditation of Educator Preparation. (2015, June 12). CAEP Accreditation Standards. Retrieved March 3, 2019, from http://caepnet.org/~/media/Files/caep/standards/final-board-amended-20150612.pdf

Freudenthal, H. (1973). *Mathematics as an educational task.* Dordrecht, The Netherlands: Reidel.

Freudenthal, H. (1991). *Revisiting mathematics education.* Dordrecht, The Netherlands: Kluwer.

Gravemeijer, K., Cobb, P., Bowers, J., & Whitenack, J. (2000). Symbolizing, modeling, and instructional design. In P. Cobb, E. Yackel, & K. McClain (Eds.), *Symbolizing and communicating in mathematics classrooms: Perspectives on discourse, tools, and instructional design* (pp. 225–274). Mahwah, NJ: Routledge.

Gravemeijer, K., & van Galen, F. (2003). Facts and algorithms as products of students' own mathematical activity. In J. Kilpatrick, W. G. Martin, & D. Schifter (Eds.), *A research companion to Principles and Standards for School Mathematics* (pp. 114–122). Reston, VA: National Council of Teachers of Mathematics.

Grossman, P., Hammerness, K., & McDonald, M. (2009). Redefining teaching, re-imagining teacher education. *Teachers and Teaching: Theory and Practice, 15*(2), 273–289.

High-Leverage Practices. (2018). *Teaching works.* Retrieved from www.teachingworks.org/work-of-teaching/high-leverage-practices

Jacobs, V. R., Lamb, L. L., & Philipp, R. A. (2010). Professional noticing of children's mathematical thinking. *Journal for Research in Mathematics Education, 41*(2), 169–202.

Lampert, M. (2009). Learning teaching in, from, and for practice: What do we mean? *Journal of Teacher Education, 61*(1–2), 21–34.

Lampert, M., Franke, M. L., Kazemi, E., Ghousseini, H., Turrou, A. C., Beasley, H., . . . Crowe, K. (2013). Keeping it complex: Using rehearsals to support novice teacher learning of ambitious teaching. *Journal of Teacher Education, 64*(3), 226–243.

Learning Mathematics for Teaching. (2005). *Mathematical knowledge for teaching measures.* Ann Arbor, MI. Retrieved from http://www.umich.edu/~lmtweb/

National Council of Teachers of Mathematics. (2014). *Principles to actions: Ensuring mathematical success for all.* Reston, VA: National Council of Teachers of Mathematics.

National Research Council. (2001). *Adding it up: Helping children learn mathematics.* Washington, DC: National Academy Press.

Saldaña, J. (2015). *The coding manual for qualitative researchers.* Thousand Oaks, CA: Sage.

Shepard, L. A. (2018). Learning progressions as tools for assessment and learning. *Applied Measurement in Education, 31*(2), 165–174.

Sherin, M. G., & van Es, E. A. (2005). Using video to support teachers' ability to notice classroom interactions. *Journal of Technology and Teacher Education, 13*(3), 475–491.

Shulman, L. S. (1987). Knowledge and teaching: Foundations of the new reform. *Harvard Educational Review, 57*(1), 1–22.

Stein, M. K., Engle, R. A., Smith, M. S., & Hughes, E. K. (2008). Orchestrating productive mathematical discussions: Five practices for helping teachers move beyond show and tell. *Mathematical Thinking & Learning, 10*(4), 313–340.

Stein, M. K., Smith, M. S., Henningsen, M., & Silver, E. A. (2000). *Implementing standards-based mathematics instruction: A casebook for professional development*. New York, NY: Teachers College Press.

Stockero, S. L. (2008). Using a video-based curriculum to develop a reflective stance in prospective mathematics teachers. *Journal of Mathematics Teacher Education, 11*(5), 373–394.

Treffers, A. (1987). *Three dimensions: A model of goal and theory description in mathematics instruction: The Wiskobas Project*. Dordrecht, The Netherlands: Reidel.

van den Kieboom, L. A., Magiera, M. T., & Moyer, J. C. (2014). Exploring the relationship between K-8 prospective teachers' algebraic thinking proficiency and the questions they pose during diagnostic algebraic thinking interviews. *Journal of Mathematics Teacher Education, 17*(5), 429–461.

Wilson, M. (2005). *Constructing measures: An item response modeling approach*. Mahwah, NJ: Lawrence Erlbaum Associates, Inc.

Appendix
Attentiveness in QRI Development Process and Instrument Summary Statement

Attentiveness in QRI Development Process

The Attentiveness in QRI was developed using a process described in Carney et al. (2017). Phase 1 involved mapping the domain of quantitative reasoning at the secondary (ages 11–17) level, including identification of mathematical tasks that target the domain and articulated progressions of how student thinking might develop within the domain. Administration constraints required us to identify exemplar quantitative reasoning tasks that highlight important ideas within the domain rather than attempt to address the entire domain of secondary quantitative reasoning.

Phase 2 involved administering the exemplar quantitative reasoning tasks to secondary student learners, analyzing student responses, and identifying exemplar student solution strategies. These exemplar student solution strategies often demonstrate less formal ways of understanding a task along a progression of student thinking.

Phase 3 involved developing constructed-response item blocks using the quantitative reasoning tasks and related student solution strategies identified in phase 2, and prompts related to (a) the mathematical intent of the task, (b) describing the student approach, (c) describing the student understanding, and (d) describing how the test taker would respond to the student based on the student work sample. These items blocks were administered to secondary mathematics teacher candidates and secondary mathematics teachers as described in the "Participants" section of the "Methods and Results." The analysis of the responses to these items is the construct-mapping process described in this chapter.

The results of the construct-mapping process can be used for phase 4 of the development process (if it is determined that a large-scale administration will require a more efficient means of scoring). Phase 4 involves the development of selected-response versions of the phase 3 assessment items. This phase includes use of selected exemplar teacher and teacher candidate responses identified during the construct-mapping process and development of an accompanying ranking scale that can be used for assessment scoring. Our phase 4 work is briefly discussed in the "Current and Future Work" section of the "Discussion."

Attentiveness in QRI Inventory Summary Statement

The intent of an instrument statement is to inform the end user as to whether instrument administration will generate score interpretations useful for their assessment needs. The elements identified as important to include when developing an instrument summary statement arose from the National Science Foundation sponsored V-M²ED conference (www.measuresinmathed.org/v-m2ed-conference). The instrument summary statement for the Attentiveness is QRI provided here describes the constructed-response version of the instrument.

The Attentiveness in Quantitative Reasoning Inventory (QRI) measures teachers' attentiveness to secondary students' quantitative reasoning. Attentiveness is the ability to analyze and respond to a particular student's mathematical ideas from a progressive formalization perspective. Attentiveness is a critical skill underlying the ability to engage in many high-leverage or "core" classroom instructional practices (Grossman et al., 2009). The inventory items are targeted at identifying teachers' and teacher candidates' ability to (a) describe the intent of a task, (b) describe the student approach, (c) interpret student understanding based on a written solution strategies, and (d) respond to a student in ways that build upon the student understanding and intent of the task within the domain of secondary students' (ages 11–17) quantitative reasoning. The Attentiveness in QRI is administered via an online survey administration system and consists of three item blocks, with each block consisting of one quantitative reasoning task, two student work samples per task, and prompts related to (a)–(d) above. The construct maps developed at the prompt level (intent, approach, understanding, and response) can be used to analyze responses and interpret scores for an individual teacher or teacher candidate as well as inform understanding of interventions designed to increase attentiveness.

8   A Validation Process for Complex Knowledge

The Standards for Mathematical Practice Knowledge Assessment

Gabriel Matney, Jonathan D. Bostic, and Matthew Lavery

Knowledge about teaching is complex on many levels (Association of Mathematics Teacher Educators [AMTE], 2017; Leonard, Brooks, Barnes-Johnson, & Berry, 2010; Jackson, Garrison, Wilson, Gibbons, & Shahan, 2013). When considering classroom teaching practice and the use of knowledge during instruction to make pedagogical decisions for students' learning, the dynamics are multifaceted (National Council of Teachers of Mathematics [NCTM], 2014). The practice of teaching mathematics involves multiple knowledge sets: knowledge of mathematics, knowledge of students, knowledge of pedagogy, knowledge of mathematical practices, knowledge of curriculum, and knowledge of social contexts (AMTE, 2017) and knowledge of mathematical and pedagogical connectedness (Matney, 2014). Because teaching draws on many knowledge types, instruments are needed to adequately measure these diverse forms of knowledge. For instance, measurement of content knowledge, pedagogical content knowledge, and classroom practices are three viable directions for measuring teachers' knowledge related to instruction. The purpose of this chapter is to share the validation process and evidence for the Standards for Mathematical Practice Knowledge Assessment (SMP-KA), which measures teachers' knowledge about the Standards for Mathematical Practice (SMP), using a lens of depth of knowledge and knowledge of standards.

Purpose Statement for Standards for Mathematical Practice Knowledge Assessment

The SMP-KA was designed to measure a teacher's knowledge of the Standards for Mathematical Practice (SMP). The SMPs are delineated in the Common Core State Standards for Mathematics (CCSSM) as eight standards elucidating what it means to be mathematically proficient (Common Core State Standards Initiative, 2010). The SMP-KA uses the Revised Bloom's Taxonomy (Anderson & Krathwohl, 2001) as a framework. A major portion of the United States' 3.1 million teachers (Common Core of Data, 2018) is in the 42 states that *expect teachers to know and use the SMPs* as part of their professional teaching obligation. The SMP-KA has four phases (see Appendix 8.A for a concise explanation of the phases) and takes between 30 and 60 minutes to administer. The target population for the SMP-KA is K-12 teachers in states that adopted the CCSSM. While the SMP-KA was developed for use with both preservice and in-service teachers, this chapter only presents and analyzes the validity evidence pertaining to its use with in-service teachers. Measurement contexts include, but are not limited, to pre- and posttesting for professional development evaluation and/or research on teacher's SMP-related knowledge. It is strongly recommended for use when considering conditions under which it might be conjectured that teachers are developing knowledge of the SMPs in robust ways, either through long-term professional development or district initiatives involving mathematical proficiency. The assessment is given via computer in which participants type their responses in a single sitting. Participants complete the four phases of the assessment one at a time and, upon completing each phase, they are locked out of going back to a prior phase. The items are not released to the public to protect the integrity of the SMP-KA. Interested users should contact the first author for pricing and use information. Only trained reviewers, who have completed a minimum of 100 hours of SMP-related professional development, score responses. Due to the extensive professional knowledge required for scoring, end users must contract with the developers for reliable scoring. The results of the scores are then returned to the end user for analysis. A total raw score may be summed across all phases of the SMP-KA for a total of 81 points (e.g., 47/81).

Theoretical Framework of Knowledge

Forty-two U.S. states have adopted the CCSSM (CCSSI, 2010) in some form as of 2018. The CCSSM include the Standards for Mathematical Practice (SMPs) and the Standards for Mathematics Content. The eight SMPs (see Table 8.1) describe mathematical behaviors and habits that students should demonstrate and teachers should seek to encourage through planning and implementation (Bostic, Matney, & Sondergeld, 2019). Because these are standards that teachers should promote during classroom instruction, it seems reasonable to infer that teachers should know what they are. Unfortunately, the rollout of new standards included insufficient funding for professional development and inadequate time to support teachers' growth as professionals (Bostic & Matney, 2013). Fast-forward to 2018 and teachers continue to experience professional development focused on the SMPs and concomitantly, preservice teachers are learning about them through university coursework and field experiences (Kruse, Schlosser, & Bostic, 2017).

Few measures are designed to reliably capture teachers' knowledge of the SMPs. The Mathematics Classroom Observation Protocol for Practices

Table 8.1 Standards for Mathematical Practice

SMP Number	Title
1	Make sense of problems and persevere in solving them.
2	Reason abstractly and quantitatively.
3	Construct viable arguments and critique the reasoning of others.
4	Model with mathematics.
5	Use appropriate tools strategically.
6	Attend to precision.
7	Look for and make use of structure.
8	Look for regularity in repeated reasoning.

Note: Discussion about a specific SMP is denoted as SMP# within the chapter.

(MCOP^2; Gleason, Livers, & Zelkowski, 2017) and the Revised Standards for Mathematical Practice Look-for Protocol (Bostic et al., 2019) allow users to connect observational data with the SMPs. While both measures are grounded effectively in robust validation evidence, there are no measures of a teachers' knowledge of the SMPs. Without such a measure, it is unclear whether there is a viable means to connect teachers' knowledge with classroom practices. Thus, there is a need to develop and validate outcomes and interpretations from a measure so that teachers' knowledge and practice can be more adequately connected and investigated.

The SMPs (CCSSI, 2010) themselves are complex notions of what it means to engage with and construct knowledge of mathematics. In the development of the Standards for Mathematical Practice Knowledge Assessment (SMP-KA), we sought a framework that would treat complex knowledge (creating and analyzing ideas) as being different from basic knowledge (e.g., recalling ideas) and that would guide the construction of items to focus on different kinds of teacher knowledge about the SMPs. The Revision of Bloom's Taxonomy (Anderson & Krathwohl, 2001) was chosen as an orienting framework for the SMP-KA due to the taxonomy's usefulness in distinguishing different types of knowledge with dual dimensionality. The taxonomy has two distinctive dimensions: cognitive process and knowledge. The cognitive process dimension has six categories: remember, understand, apply, analyze, evaluate, and create. The knowledge dimension has four categories: factual, conceptual, procedural, and metacognitive. These categories are thought to vary in complexity, and as such form a hierarchy (Krathwohl, 2002). Taken collectively, these two dimensions form a matrix of 24 possibilities for classifying knowledge. For instance, knowledge might be considered to fit within the analyze and conceptual cell of the matrix, and could be described as "analyze conceptual knowledge" (ACK).

The SMP-KA does not measure all knowledge a teacher might have about the SMPs. The limitations of a survey inhibit what can be revealed about actual instruction, such as how teachers' use their knowledge of

SMPs to plan and enact instruction. The field has done well to develop a few instruments with validity evidence that measure teachers' use of SMPs within observable instruction (see Bostic et al., 2019; Gleason et al., 2017). These instruments allow researchers to gather evidence of the execution and implementation of knowledge, both of which are associated with the cognitive dimension of application in the Revised Taxonomy (Anderson & Krathwohl, 2001). What the field lacks however, is a critical examination of the other knowledge categories such as factual knowledge and conceptual knowledge; the SMP-KA fills much of this needed gap. This chapter provides a validity argument for outcome interpretations from the SMP-KA's assessment of teachers' knowledge in the areas of Remembering Factual Knowledge (RFK), Understanding Conceptual Knowledge (UCK), Analyzing Conceptual Knowledge (ACK), Creating Factual Knowledge (CFK), and Creating Conceptual Knowledge (CCK).

Instrument Validation Process

We drew upon our own previous work and the work of others that addresses validating outcomes from measures and assessments (American Educational Research Association, American Psychological Association, National Council on Measurement in Education, 2014; Bostic et al., 2019; Bostic, Sondergeld, Folger, & Kruse, 2017). Table 8.2 lists the action steps as well as their connections to validity. Following Table 8.2 are detailed descriptions of the procedures and analysis for validating outcomes.

Stage One

In considering literature related to the SMPs, we began by analyzing the SMP paragraphs from the CCSSM (CCSSI, 2010, see pp. 6–8). We read each SMP four times using different lenses. Upon the first read through, we looked for the keywords that illuminated the big picture of the SMP. The second read through focused on separating mathematical examples from more general statements about the SMP. On the third read through, we considered how the examples for each SMP connected to the general statements. This was done for the purpose of considering future examples that would honor the intent of the SMP. The fourth read through was done to find explicit connections across all eight SMPs.

After giving focused attention to understanding the SMPs, an examination of literature written by education experts since 2010 occurred via a review of manuscripts, proceedings, and presentations about SMPs. In addition to research literature, other expert ideas about the SMPs were sought from nationally known groups that have profound influence in mathematics education (e.g., NCTM and the National Council

Table 8.2 Alignment of Stages and Actions for Validating the SMP-KA

Stage	Description of Stage	Actions Completed During This Study	Source of Validity Evidence
1	Literature review.	Examined other SMP protocols, reviewed literature on SMPs, knowledge, and MKT.	Test content
2	Conduct interviews with content experts and potential tool users to consider ideas on knowledge of SMPs.	Conducted interviews with an expert panel consisting of K-12 math teachers, math coaches, mathematicians, mathematics teacher educators, and a mathematics curriculum coordinator and a state department of education mathematics representative.	Test content
3	Synthesize data from literature review and interviews with content experts to discern relevant knowledge types.	Employed typological analysis to generate five levels of knowledge to be assessed; Remember Factual SMP Knowledge, Understand Conceptual SMP Knowledge, Analyze Conceptual SMP Knowledge, and Creating Factual and Conceptual SMP Knowledge.	Internal structure
4	Item development.	Created items for each of the five knowledge categories.	—
5	Expert panel review.	Submitted items of the Standards for Mathematical Practice Knowledge Assessment to expert panel.	Test content
6	Conduct interviews with potential users of tool and synthesize from these data.	Conducted one-on-one and small-group cognitive interviews with K-12 math teachers, a curriculum coach, a curriculum coordinator, and a mathematics educator.	Response processes
7	Pilot testing the assessment.	Collected 285 instances of SMP-KA results completed by 189 participants (approximately 50% of sample completed both a pre and post).	Internal structure and evidence of reliability and internal consistency
8	Conducting psychometric analysis of collected assessment data.	Performed exploratory factor analysis and calculated inter-rater reliability.	Internal structure and evidence of reliability and internal consistency

of Supervisors of Mathematics). This literature included books such as *Connecting the NCTM Process Standards and the CCSSM Practices* (Koestler, Felton, Bieda, & Otten, 2013) and *Principles to Action: Ensuring Mathematical Success for All* (NCTM, 2014). The literature provided insights about the significance of the ideas found in the SMPs, specifically what knowledge teachers need to promote the SMPs and why it is important for teachers to have knowledge of the SMPs. While the action steps completed at this stage did not provide validity evidence to support the intended use of the SMP-KA, it was an important first step in understanding the depth and breadth of the knowledge of the SMPs that the instrument was being designed to measure, allowing developers to collect appropriate validity evidence based on test content during stage five (American Educational Research Association, American Psychological Association & National Council on Measurement in Education, 2014).

Stage Two

An expert panel consisting of 31 individuals was created to ascertain data about knowledge of the SMPs. All members of the panel were selected because of their extensive work involving the SMPs (nature of the involvement described below) and their intended promotion of the SMPs through their respective positions. The representatively selected expert panel included mathematics teaching professionals from nine different perspectives: K–5, 6–8, and 9–12 mathematics teachers; mathematics instructional coaches; mathematicians; mathematics education graduate students; mathematics educators; a mathematics curriculum coordinator; and a state department of education mathematics representative.

K-12 mathematics teachers participating in the expert panel had completed more than 120 hours of SMP focused on professional development. Each teacher had worked to promote student engagement in the SMPs and previously had shown instruction that met or exceeded norms on the Revised Standards for Mathematical Practice Look-for Protocol (Bostic et al., 2019; Bostic & Matney, 2014). The teachers' professional experience ranged from 5 to 28 years. There were six teachers from grades K–5, six teachers from grades 6–8, and six teachers from grades 9–12, which resulted in a total of 18 teachers. There was at least one teacher from each grade level on the expert panel. We sought input from two mathematics coaches, a curriculum coordinator, and a state department of education representative. These experts work at the district or state level to support mathematics teachers in the promotion of the SMPs during instruction. We reached out to obtain feedback from two mathematicians who held a terminal degree (Ph.D.). The mathematicians were part of professional development teams who worked with K-12 mathematics teachers on promoting the SMPs during instruction. They have taught a variety of graduate mathematics courses and courses for K-12 mathematics teachers and preservice teachers. Four mathematics education graduate students provided input based on their work to create, enact, and research grant-funded professional development about SMP instruction. Lastly, three mathematics teacher educators holding Ph.D.s, and hailing from different states, were asked to serve on the panel. These experts were selected due to their research involving SMPs, which had been presented at peer-refereed national mathematics education conferences and/or published in peer-reviewed mathematics education journals. During this stage, members of the expert panel communicated with instrument developers in face-to-face meetings, telephone calls, and email. Each communication was done individually and no expert panel member had knowledge or influence on the others' responses. For developers to fully understand the knowledge content that must be measured by the SMP-KA, developers asked the panel to respond to the following questions:

- What knowledge of the SMPs are useful in your own work?
- How much do teachers you work with know about the SMPs? What kinds of things do they know? The titles, examples, other things?
- What should teachers know about the SMPs in order to be effective at promoting the SMPs through their teaching with the students?

The panel provided important data about which SMPs teachers may find difficult to understand. For example, nearly all of the panel members described both their own initial difficulty, as well as colleagues' difficulties, in making sense of how SMP7 and SMP8 are alike and different. For many teachers, there was a subtle nuanced difference between students noticing a pattern or structure in SMP7 and students recognizing repeated reasoning in SMP8. The panel also shared examples of misconceptions that teachers might have. The most common misconception discussed by the panel was that some teachers confuse the strategic use of manipulatives by students to solve a problem (SMP5) as necessarily meaning that students were modeling with mathematics (SMP4). As the word "model" holds many meanings in the English language, such confusions are perhaps quite natural at first and suggest that teachers need opportunities to become more knowledgeable about the meaning of the SMPs. Other important data the panel provided involved the connections between the SMPs themselves. Teachers who know about the SMPs will see connections between them, such as pausing in the process to consider the contextual meaning (SMP2) and the behavior of students who "routinely interpret mathematical results in the context of the situation and reflect on whether the results make sense" (SMP4; CCSSI, 2010, p. 7). Teachers knowledgeable about the SMPs might then further note that this kind of contextual mathematical thinking is similar to students being able to "maintain oversight of the process, while attending to the details" (SMP8; CCSSI, 2010, p. 8). The ideas presented by the expert panel were analyzed in stage three to develop item types for an assessment that would capture data about teachers' knowledge of the SMPs. In addition, similar to stage one, the action steps completed during stage two provided a different perspective on knowledge of the SMPs for collection of validity evidence based on test content at a later stage.

Stage Three

In this stage, typological analysis was used to systematically analyze data drawn from the literature review and expert panel (Hatch, 2002). In typological analysis, data are divided into categories based on a predetermined framework (LeCompte & Preissle, 1993). For the SMP-KA, the Revised Taxonomy (Anderson & Krathwohl, 2001) was used to consider what types of items should be included in the instrument to ensure appropriate measurement of the different kinds of knowledge and cognitive process dimensions involving the SMPs. The research team considered all 24 cross-dimensional possibilities of the Revised Taxonomy.

First, the data were read and memos were made connecting the evidence from the literature and expert panel's statements concerning the types of important knowledge for SMPs. In the second step, the main ideas in each of the identified knowledge types were recorded. Step three was to reconsider the data with a focus on each of the identified knowledge types to ensure that all codes associated with each knowledge type were documented.

In step four, the research team sought relationships within each typology. The aim of this step was to consider which knowledge types had viable evidence from the literature and/or ideas from the expert panel warranting their inclusion in the SMP-KA. For example, in order for the knowledge type Analyzing Conceptual Knowledge to be viable within the SMP-KA, there must be evidence from the data that it is important for teachers to be able to break down the SMPs, detecting how they relate to one another having knowledge of "the interrelationships among the basic elements" found within the "larger structure that enables them to function together" (Krathwohl, 2002, p. 214). From such analyses, five viable knowledge types emerged: Remembering Factual Knowledge (RFK), Understanding Conceptual Knowledge (UCK), Analyzing Conceptual Knowledge (ACK), Creating Factual Knowledge (CFK), and Creating Conceptual Knowledge (CCK). In step five, the literature and ideas from the expert panel were investigated again for evidence that would counter these five knowledge types. Although no counter evidence was found, we noted that one mathematics coach mentioned, "When an expert teacher knows how to really engage students in becoming mathematical thinkers, I wonder whether or not they can do that, even if they have no idea what the SMP titles are? It seems like they could. So, while knowing the titles could be beneficial I'm not sure it's necessary." The idea here that teachers could know about habits of effective mathematical thinking without knowing the factual titles of the SMPs is important. The mathematics coach is pointing out that a teacher might be able to demonstrate that they understand the SMPs conceptually (UCK) even if they cannot recite the titles of the SMPs (RFK). Thus, the research team was alerted to carefully consider the formation of items and phases (stage 4) that allowed teachers to show their knowledge of RFK independently of other knowledge types and vice-versa.

In the final step, we selected data excerpts that supported the emergence of the five knowledge types. These data excerpts were used to guide item construction for each of the five knowledge types. See Appendix 8.B for exemplary evidence related to the five knowledge types. Similar to stages one and two described above, stage three did not provide validity evidence to support the proposed use of the SMP-KA, per se. Instead, this stage provided developers with a thorough understanding of how the Revised Taxonomy (Anderson & Krathwohl, 2001) applies to knowledge of the SMPs. Thus, when validity evidence related to internal structure (as described in American Educational Research Association et al., 2014) was collected in stages seven and eight, it could be compared to the theory which informed development of the SMP-KA.

Stage Four

Drawing upon results of the typological analysis in stage three, items were created for each of the five knowledge types. For RFK and UCK items, lists of factual and conceptual knowledge were fashioned from the SMP titles, the literature, and the SMP paragraphs. These lists consisted of phrases or sentences describing a focused aspect of the SMP. For example, two aspects associated with SMP1 would be that students look for entry points to a problem's solution and students do not give up after the first attempt. Next, each aspect was analyzed by cross referencing the aspect to the knowledge type, RFK or UCK. These aspects formed the basis for items about teachers' factual knowledge (RFK) and conceptual understanding of the SMPs (UCK).

For the development of items associated with ACK, CFK, and CCK, written exemplary scenarios of students engaging in the SMPs while solving mathematics problems were developed. The scenarios were created from actual events with K-12 students who were solving mathematics problems. They were taken from both live observations and videos ($n = 591$) of teaching where the teacher's lesson plan had an explicit focus on the promotion of at least one SMP. The videos of classroom instruction came from K-12 teachers participating in a three-year professional development program focused on SMPs. The purpose of developing these scenarios was to draw upon genuine classroom happenings to capture students' engagement in the SMPs while doing specific mathematics problems. Both the process and the product of writing SMP-focused scenarios formed the basis for items about analyzing SMPs (ACK) and creating scenarios of SMPs that demonstrate factual and conceptual knowledge (CFK and CCK).

After the construction of the items, the assessment was organized into four phases (see Appendix 8.A). The phases were then ordered in a way that would not reveal ideas about the SMPs before the teachers completed each phase. Every participant completed phase 1, phase 2, phase 3, and finally phase 4, in that order, and once a phase was completed teachers could not go back to a previous phase. For example, the ordering does not allow teachers to see an exemplar scenario and then analyze it before they are asked to create one of their own, because doing so might attune teachers to what is possible and obfuscate assessment of their prior knowledge of the SMPs. Table 8.3 shows the alignment and order of each section of the SMP-KA with its associated knowledge type. Three of the phases (1, 2, and 4) are each associated with exactly one knowledge type. Phase 3 is associated with two knowledge types. Krathwohl (2002) explains that particular items and objects will sometimes require participants to engage in more than one type of knowledge. In the case of phase 3, the nature of the task requires participants to create factual knowledge and create conceptual knowledge (see Appendix 8.A for more detail). Developing items that indicate what a teacher knows about the SMPs requires thoughtful consideration. Close attention was paid to what was written in the SMPs (CCSSI, 2010) as well as the ways the mathematics education literature and experts described the SMPs. Furthermore, we looked at actual classroom happenings, via the videos mentioned previously, to ensure the items fit within the contexts of student's mathematical problem solving.

Stage Five

A goal of this stage was to collect validity evidence based on test content. Such evidence connects the content within an instrument and the

Table 8.3 SMP-KA Knowledge Alignment and Order of Administration

Phase	Knowledge Type
1	RFK
2	UCK
3	CFK and CCK
4	ACK

Note: RFK = Remembering Factual Knowledge, UCK = Understanding Conceptual Knowledge, CFK = Creating Factual Knowledge, CCK = Creating Conceptual Knowledge, ACK = Analyzing Conceptual Knowledge. Respondents must complete each phase before seeing the items in the next phase.

construct it intends to measure (American Educational Research Association et al., 2014) and, in turn, provides assurances that the score interpretations are appropriately drawn. To collect validity evidence based on test content, an expert panel of 13 individuals reviewed the assessment developed from the previous stage. It consisted of a representative sample of individuals from each group described earlier in stage two; two terminally degreed mathematicians, two terminally degreed mathematics educators, three teachers from each grade band (K-5, 6–8, 9–12), three mathematics education graduate assistants, two mathematics coaches from elementary (K-8) and secondary (7–12), and one mathematics curriculum coordinator. The panel examined the SMP-KA and gave verbal or written feedback about the connections they made between the items and knowledge of the SMPs. Panel members were asked to directly respond to the appropriateness of the items in representing a teacher's knowledge about the SMPs. This was done for each knowledge type (RFK, UCK, CFK, CCK, and ACK). The feedback was positive and suggestions for improvement were shared. The assessment was modified to include connections between the SMPs where they were warranted and some of the exemplar scenarios were eliminated based on the experts' feedback. Including all of the scenarios made the assessment very long. The consensus of the panel was that one exemplar scenario per SMP was sufficient. Eight scenarios were selected that the panel agreed were representative of the eight SMPs. In other words, for each SMP there was exactly one matching scenario. These modifications were made to the instrument before collecting data during stage six.

Stage Six

Cognitive interviews can be used to improve the validity and reliability of research tools (Desimone & Le Floch, 2004) and to provide validity evidence based on response processes (American Educational Research Association et al., 2014; Castillo-Díaz & Padilla, 2013). One-on-one cognitive interviews were conducted with participants who would be expected to take the assessment (preservice and in-service teachers) and potential users of the assessment results. There were a total of 15 participants during this stage: 6 preservice teachers, 6 in-service teachers, 1 mathematics coach, 1 mathematics curriculum coordinator, and 1 terminally degreed mathematics educator. The focus of the cognitive interviews was to inquire about the meaning of the items by user and to gauge the assessment's ease of use via the computer. The cognitive interviews were transcribed and inductive analysis (Hatch, 2002) was conducted to determine any themes or discrepancies in meanings of items across potential assessment takers and assessment users. The results of our analyses revealed that all phases of the assessment were interpreted as intended and easy to use via the computer; except one. These interviews rewarded us with important information about how the directions for phase 2 were being interpreted. The problematic interpretation of the directions did not elicit the desired information about teachers' knowledge of the SMPs. The directions for phase 2 were reworded to be more direct and clearer for participants. After rewording the directions, the teachers gave feedback about the expectations of the directions that aligned with the intent. This change was made prior to stage seven.

Stage Seven

In this stage, we piloted the assessment with K-12 teachers from a Midwest state that adopted the CCSSM in 2011, five full years before teacher participants joined this validation study. Teachers who participated in the pilot were enrolled in a grant that included professional development, which had the study of the SMPs as one of its stated goals. The pilot program included more than 68 hours of professional development that either directly examined the descriptions of the SMPs (CCSSI, 2010) or explored the SMPs through the context of problem solving and student engagement in the Standards for Mathematics Content. Teachers took the assessment on two occasions. The first administration was prior to any grant-related professional development and then again after the conclusion of the grant. The teachers came from rural, suburban, and urban schools. Most of the teachers were female and Caucasian. During this stage, 189 participants completed the SMP-KA 285 times, with half of the participants (n = 94/189, 49.7%) completing the assessment both before and after the grant activities. The pilot administration of the SMP-KA at this stage provided the data that was analyzed during the next stage to provide validity evidence based on internal structure (American Educational Research Association et al., 2014) and evidence of the instrument's reliability and internal consistency.

Stage Eight

We performed reliability analyses and exploratory factor analyses to determine the psychometric properties and performance of the SMP-KA. As a first step to determine instrument reliability, inter-rater agreement was calculated for five raters. Due to the complexity of SMP knowledge, only raters who had more than 100 hours of SMP professional development were allowed. Each rater underwent about 10 additional hours of training. The training process was done in teams of two or more and began with two hours of SMP-KA rubric reading followed by making explicit connections between the rubric, the SMP paragraphs (CCSSI, 2010), and literature. Next, the raters were given data from five participants. These data were chosen to show the breadth, depth, and variance of teachers' answers. The raters then scored the first participants' data together with the first author allowing for discussion on the meanings of the participants' responses and how to code consistently. Next, the raters scored four more participants' data independently. After independent scoring, the raters reconvened with the first author to discuss each score. Any discrepancies between the scores were discussed until agreement was made. After this training and calibration experience, each rater scored six new assessments independently. These independent assessments were then analyzed and revealed 97.1% exact agreement across coders. This exceeds the minimum threshold of 90% needed to conduct reliability analyses (James, Demaree, & Wolf, 1993), and represents a stronger indicator of consistency than inter-rater reliability insofar as it suggests that raters interpret participant responses similarly and assign codes virtually identically (Gall, Gall, & Borg, 2007).

Exploratory factor analyses (EFA) were used to understand the underlying structure of the instrument. EFA was appropriate in this case, not only because these items were just developed and had not been analyzed together previously (Bandalos & Finney, 2010), but also because of the theory on which they were designed. The instrument was developed to conform to two different theoretical structures simultaneously; each phase of the assessment measured different dimensions of knowledge according to the Revised Taxonomy (Anderson & Krathwohl, 2001), and each phase also contained items about each of the SMPs (CCSSI, 2010). As such, it was reasonable to use EFA rather than confirmatory factor analysis (CFA), as it is unclear to which theoretical framework the emergent subfactors would align (Bandalos & Finney, 2010). The SMP-KA measures teachers' deep and complex knowledge of the SMPs, which was conceptualized as a single higher-order factor that incorporated all of the information contained in the subfactors that might emerge. Results of the EFA offered insight into the psychometric qualities of the SMP-KA and allowed researchers to interpret subscale scores to examine distinct, though related, aspects of the knowledge it measures. We conducted EFA using maximum likelihood estimation in SPSS Statistics, version 24 (IBM Corp., 2016). Because factors were expected to correlate, we selected an oblique rotated solution using the Promax method with Kaiser normalization (Field, 2018). Four factors emerged with eigenvalues greater than one. Inspection of the scree plot (see Figure 8.1) revealed an inflection point at four factors, confirming the four-factor solution (Field, 2018).

Each identified factor contained the items included in each phase of the assessment and jointly explained 67.0% of the total variance in participant scores before rotation. After rotation, factors overlap, preventing calculation of total variance explained (Bandalos & Finney, 2010), but the sums of squared structure coefficients for each factor can be used to compare the amount of variance uniquely explained by each factor. Table 8.4 displays the means, standard deviations, pattern coefficients

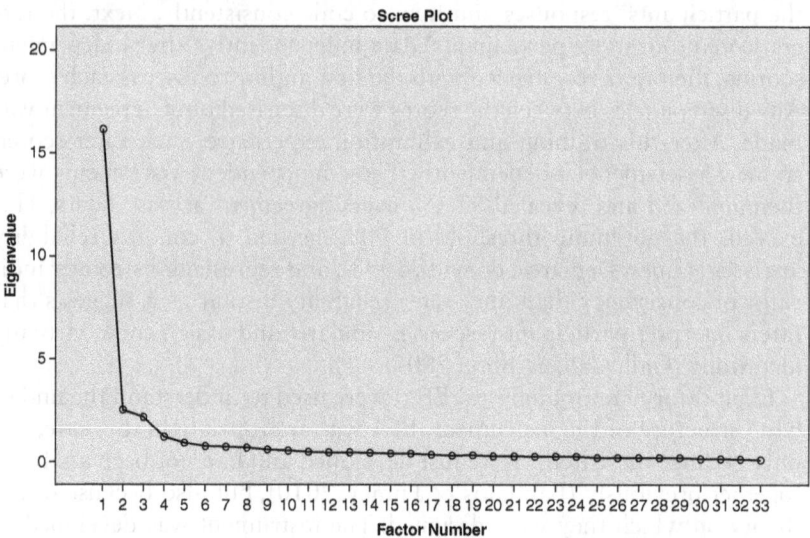

Figure 8.1 Scree plot from exploratory factor analysis conducted during stage eight of the SMP-KA validation process.

(which represent the unique relationship between the item and the underlying factor after controlling for the other factors), and communalities for each item in the SMP-KA, as well as the phase, factor, sum of squared structure coefficients, and Cronbach's alpha value (discussed next) for each grouping of items.

Analysis of the internal consistency for each subscale using Cronbach's alpha indicates reliabilities either above .90 or, in the case of the CFK/CCK factor, very near it, and are suitable for the purposes of this instrument (Lance, Butts, & Michels, 2006). The overall scale that included all items in the instrument demonstrates high reliability with an internal consistency of $\alpha = .97$, which is excellent (Lance et al., 2006). Internal consistency analyses for the overall scale, and for each subscale, indicated that dropping any of the items associated with that scale would not improve the internal consistency, suggesting that each item contributes well to the measure of both teachers' complex content knowledge of the SMPs and its corresponding knowledge dimension. Factor correlations (see Table 8.5) suggest that the use of oblique rotation was appropriate, as all four factors are highly correlated (Cohen, 1992). These high correlations also suggest that, although the scores for each phase can provide information on its associated knowledge dimension, the results of the SMP-KA are best interpreted as a single total score representing a teacher's complex knowledge of the SMPs.

Table 8.4 Means, Standard Deviations, Pattern Coefficients, and Communalities for SMP-KA Items

Pattern Coefficient

Item	M	SD	RFK	UCK	CFK/CCK	ACK	h^2
Phase 1, RFK Factor (SSSC = 12.2, Cronbach's α = 0.95)							
Q1A	0.69	0.83	**0.81**	−0.04	0.02	0.06	0.71
Q1B	0.62	0.85	**0.90**	0.04	−0.12	0.05	0.79
Q1C	0.49	0.79	**0.88**	0.02	−0.13	0.10	0.82
Q1D	0.37	0.49	**0.84**	−0.10	0.01	0.01	0.66
Q1E	0.50	0.76	**0.79**	0.03	0.05	0.08	0.79
Q1F	0.32	0.48	**0.81**	−0.06	0.05	0.05	0.73
Q1G	0.50	0.83	**0.84**	0.07	0.07	−0.05	0.80
Q1H	0.31	0.70	**0.80**	0.09	0.10	−0.19	0.71
Phase 2, UCK Factor (SSSC = 12.3, Cronbach's α = 0.94)							
Q2	0.60	0.90	−0.04	**0.66**	0.11	0.11	0.67
Q3	0.36	0.73	0.12	**0.68**	0.02	0.15	0.81
Q4	0.47	0.77	0.04	**0.68**	−0.04	0.07	0.61
Q5	0.27	0.65	0.12	**0.88**	0.11	−0.19	0.79
Q6	0.36	0.64	−0.08	**0.73**	−0.09	0.12	0.54
Q7	0.34	0.69	−0.14	**0.82**	0.05	0.11	0.74
Q8	0.16	0.51	0.06	**0.81**	−0.10	0.00	0.71
Q9	0.19	0.57	0.01	**1.02**[a]	−0.04	−0.22	0.73
Phase 3, ACK Factor (SSSC = 13.1, Cronbach's α = 0.94)							
Q10	0.44	0.73	−0.10	0.10	0.07	**0.76**	0.70
Q11	0.40	0.81	0.10	−0.13	0.00	**0.76**	0.62
Q12	0.42	0.81	0.03	−0.03	−0.06	**0.79**	0.64
Q13	0.58	0.80	−0.03	0.04	0.19	**0.68**	0.70
Q14	0.39	0.74	0.11	0.10	−0.05	**0.64**	0.60
Q15	0.51	0.81	0.03	−0.01	0.13	**0.76**	0.74
Q16	0.43	0.76	0.06	0.22	0.08	**0.57**	0.74
Q17	0.40	0.78	0.11	0.13	0.26	**0.35**	0.62
Q18	0.40	0.82	0.13	0.29	0.14	**0.36**	0.71
Phase 4, CFK/CCK Factor (SSSC = 10.7, Cronbach's α = 0.87)							
Q19	0.65	0.80	0.05	0.02	**0.67**	0.01	0.52
Q20	0.33	0.60	0.04	0.03	**0.47**	0.09	0.42
Q21	0.79	0.86	−0.01	0.02	**0.84**	−0.06	0.63
Q22	0.29	0.61	−0.17	0.13	**0.44**	0.13	0.39
Q23	0.57	0.81	0.02	0.12	**0.64**	−0.12	0.46
Q24	0.43	0.68	0.13	−0.21	**0.55**	0.13	0.47
Q25	0.87	0.88	−0.10	−0.14	**0.99**	−0.01	0.66
Q26	0.40	0.71	0.17	0.07	**0.54**	−0.04	0.49

Note: Bold indicates highest pattern coefficient. h^2 = communality, SSSC = sum of squared structure coefficients.

[a] In an EFA with an oblique rotated solution, pattern coefficients are analogous to beta weights in multiple regression analyses and can fall outside the +1 to −1 range, unlike correlation coefficients.

Table 8.5 Factor Correlations

Factor	1	2	3	4
RFK	—			
UCK	.64	—		
CFK/CCK	.55	.58	—	
ACK	.68	.71	.73	—

Discussion

The SMP-KA was carefully designed to address a gap in the measurement of teachers' knowledge of the SPMs. Using the Revised Taxonomy (Anderson & Krathwohl, 2001) as a framework, the SMP-KA accounts for measures of teacher's knowledge within five domains (RFK, UCK, ACK, CFK, and CCK) that were previously unmeasured. Furthermore, the SMP-KA gives a total score by which an increase of these knowledge domains can be measured as a teacher's knowledge of the SMPs expands. These data may interest schools and districts who want to know about teachers' knowledge of the SMPs, in order to make data-based decisions on future professional development, analyze district compliance with implementation of the standards, measure the level of preservice or in-service teacher knowledge of SMPs, or consider how teacher knowledge connects with their students' mathematical behaviors and actions. Similarly, researchers wanting to investigate teachers' knowledge of the SMPs could employ the SMP-KA in order to find out the effects of SMP-related professional development or other correlated variables.

The development of the SMP-KA was done to complement the observational protocols related to the SMPs (see Bostic et al., 2019; Gleason et al., 2017). These observation protocols may include data for assessing teacher's knowledge of specific mathematics skills and algorithms, implementation of techniques and pedagogical procedures related to promotion of the SMPs, and knowledge of when to apply appropriate mathematical and pedagogical procedures. Classroom observation data have the power to reveal teacher's execution and implementation of the SMPs, and hence are particularly well suited to give robust information about teacher's SMP knowledge in the domains of application of factual knowledge, application of conceptual knowledge, and more broadly in teachers' remembering, understanding, applying, analyzing, and evaluating procedural knowledge of the SMPs. On the other hand, these instruments do not consider other areas of knowledge, and it was within these spaces that our development of the SMP-KA sought to inquire about other areas of knowledge measurement would be valuable to the field. The expert panel revealed five untapped areas of knowledge (RFK, UCK, ACK, CFK, CCK) for the SMP-KA to assess. It seems plausible then, that those looking to have a large swath of information related to teachers' knowledge of the SMPs could couple the SMP-KA with classroom observation protocols, in order to gain a strong picture of teachers' SMP knowledge across several domains.

Participants in this study worked through the items on the SMP-KA at different rates, ranging from 30 to 60 minutes. Based on these analyses and on the idea that the eight SMPs constitute a single, overarching psychological construct that includes a variety of diverse mathematical behaviors and habits, it is plausible that a short form of the SMP-KA can be developed, which would provide similar measurement quality while requiring less time to complete the assessment. Such a short form must be systematically constructed with careful attention to the EFA and internal consistency analyses and to align with theory.

Future Research and Limitations

As mentioned in the discussion section, the SMP-KA is a measure of teachers' SMP knowledge and is not a measure of enacted instruction. Teaching mathematics is a complex endeavor (AMTE, 2017) in which teacher's use an array of knowledge in their planning and in the moment-by-moment orchestration of the student learning. Although knowledge plays an important role in all elements of teaching, consideration of a teacher's knowledge, by itself, limits teachers, school leaders, researchers, and policymakers' abilities to make decisions about what is needed to improve teaching and learning. Thus, future research should explore evidence connecting knowledge of the SMPs with enacted SMP instruction. Similarly, the SMP-KA may or may not show correlation to student outcome measures, including but not limited to, achievement, problem solving, and affect.

The field of mathematics education has long developed knowledge measures for teachers in various knowledge domains, such as the DTAMS (Saderholm, Ronau, Brown, & Collins, 2010) and LMT (Ball, Thames, & Phelps, 2008) for mathematical knowledge for teaching. Researchers have also examined under what professional development conditions these measures tend to perform better (Copur-Gencturk & Lubienski, 2013). The SMP-KA provides a direct and focused measure of teachers' knowledge of the SMPs. Future research should explore how teachers' knowledge of the SMPs connect with other knowledge measures that are frequently used in the field as well as the professional development conditions under which teachers' knowledge of SMPs provides substantive benefits.

At this time, all of the work done with the SMP-KA has been by researchers and mathematics education graduate students who were part of the development team. Furthermore, the participants for the validity examination came from professional development programs created by the developers in the years prior to its development. For these reasons, further research is needed to see what happens when others use the SMP-KA.

Final Thoughts

This chapter provides validity evidence for the score interpretations from the SMP-KA as a measure of in-service teacher's complex knowledge of the SMPs, and connects that evidence to the measure development process. Table 8.6 concisely shows the sources of evidence, the validity claims, and supporting evidence given in this chapter.

As shared in Table 8.6, this is not the only time that validity evidence will be gathered because validation is an ongoing process (American

Table 8.6 Validity Argument for the Interpretation of the SMP-KA as a Measure of Inservice Teacher's Complex Knowledge of the SMPs

Source of Validity Evidence	Claim	Evidence
Test Content	• The knowledge of the SMPs required for effective mathematics instruction is complex and multidimensional	• Confirmed by systematic review of literature before development • Confirmed through interviews with practitioners and experts before development
	• The SMP-KA adequately measures the complexity and multidimensionality of SMP knowledge	• Confirmed through expert review of the SMP-KA before piloting
Response Processes	• The SMP-KA's phase structure engages respondents in the appropriate response processes to measure the intended knowledge domain (i.e., the intended knowledge dimension and the intended cognitive process dimension)	• Confirmed through cognitive interviews with a representative sample of respondents
Internal Structure	• The SMP-KA measures the intended knowledge domain independently of the other knowledge domains	• Confirmed by exploratory factor analysis results
Relations to Other Variables	--	Currently being collected
Related Consequences	--	Currently being collected

Note: Table developed based on the recommendations of Ferrara, S. (2007). Our field needs a framework to guide development of validity research agendas and identification of validity research questions and threats to validity. *Measurement: Interdisciplinary Research and Perspectives, 5*(2–3), 156–164. doi:10.1080/15366360701487500. Sources of validity evidence are derived from the Standards for Educational and Psychological Testing (American Educational Research Association, American Psychological Association, & National Council on Measurement in Education, 2014).

Educational Research Association et al., 2014; Bostic, Krupa, Carney, & Shih, this volume; Kane, 2012). For example, validity evidence data need to be gathered from preservice teachers to explore how their score interpretations are similar to or different from in-service teachers. Another opportunity for further research involves gathering validity evidence in relation to other variables by exploring connections between measures of pedagogical content knowledge (e.g., DTAMS and LMT) and the SMP-KA. We will continue to explore ways to strengthen the validity argument for this measure of teachers' knowledge of the SMPs.

The SMP-KA provides a means to measure teachers' knowledge of the SMPs. As such, schools, districts, or states in which the CCSSM was adopted may use the SMP-KA as a gauge for where teachers are in their knowledge of mathematical proficiency. Those providing professional development involving learning about SMPs, or what mathematical proficiency means in the CCSSM, might use the SMP-KA to gather evidence of changes in teacher knowledge. Researchers who are interested in the relationship between teachers' knowledge and instruction may use the SMP-KA as one tool in studies exploring the enacted dynamics of teachers' knowledge about mathematical proficiency. With that said, the SMP-KA provides scholars with a means to explore teachers' knowledge of the SMPs connected to numerous contexts. There is sufficient validity evidence related to a variety of sources necessary to ground the score interpretations, when the SMP-KA is used appropriately. And in sum, this chapter may help readers who are thinking about how the measure development process and validity gathering process occur in tandem with one another.

Author Note

This research was supported in part by a grant from the Ohio Department of Education (Grant #016437). Any opinions, findings, conclusions, or recommendations expressed by the authors do not necessarily reflect the views of the Ohio Department of Education.

We would like to recognize the following individuals for their efforts in the development of the SMP-KA: Jacob Burgoon, Corrinne Fischer, Alyssa Lustgarten, Julia Porcella, Allison Marino, Megan Schlosser, and Rachel Weimken.

References

American Educational Research Association, American Psychological Association, & National Council on Measurement in Education. (2014). *Standards for educational and psychological testing.* Washington, DC: American Educational Research Association.

Anderson, L. W., & Krathwohl, D. R. (Eds.). (2001). *A taxonomy for learning, teaching, and assessing: A revision of Bloom's Taxonomy of Educational Objectives*. New York: Longman.

Association of Mathematics Teacher Educators. (2017). *Standards for preparing teachers of mathematics*. Retrieved from https://amte.net/sptm

Ball, D. L., Thames, M. H., & Phelps, G. (2008). Content knowledge for teaching: What makes it special? *Journal of Teacher Education, 59*(5), 389–407. doi:10.1177/0022487108324554

Bandalos, D. L., & Finney, S. J. (2010). Factor analysis. In G. R. Hancock & R. O. Mueller (Eds.), *The reviewer's guide to quantitative methods in the social sciences*. New York, NY: Taylor & Francis.

Bostic, J., Krupa, E., Carney, M., & Shih, J. (2019). Reflecting on the past and thinking ahead in the measurement of students' outcomes. In J. Bostic, E. Krupa, & J. Shih (Eds.), *Quantitative measures of mathematical knowledge: Researching instruments and perspectives*. New York, NY: Routledge.

Bostic, J., & Matney, G. (2013, March). Preparing K-10 teachers through common core for reasoning and sense making. In S. Reeder & G. Matney (Eds.), *Proceedings of the 40th annual meeting of the Research Council on Mathematics Learning*. Tulsa, OK. Retrieved from http://web.unlv.edu/RCML/2013 Proceedings.pdf

Bostic, J., & Matney, G. (2014). Role-playing the standards for mathematical practice: A professional development tool. *Journal for Mathematics Education Leadership, 15*(2), 3–10.

Bostic, J., Matney, G., & Sondergeld, T. (2019). A lens on teachers' promotion of the Standards for Mathematical Practice. *Investigations in Mathematics Learning, 11*(1), 69–82. doi.org/10.1080/19477503.2017.1379894

Bostic, J., Sondergeld, T., Folger, T., & Kruse, L. (2017). PSM7 and PSM8: Validating two problem-solving measures. *Journal of Applied Measurement, 18*(2), 151–162.

Castillo-Díaz, M., & Padilla, J.-L. (2013). How cognitive interviewing can provide validity evidence of the response processes to scale items. *Social Indicators Research, 114*(3), 963–975. doi:10.1007/s11205-012-0184-8

Cohen, J. (1992). A power primer. *Psychological Bulletin, 112*(1), 155–159. doi:10.1037/0033-2909.112.1.155

Common Core of Data. (2018). Retrieved from https://nces.ed.gov/ccd/pub_snf_report.asp

Common Core State Standards Initiative. (2010). *Common core standards for mathematics*. Washington, DC: National Governors Association Center for Best Practices and Council of Chief State School Officers.

Copur-Gencturk, Y., & Lubienski, S. T. (2013). Measuring mathematical knowledge for teaching: A longitudinal study using two measures. *Journal of Mathematics Teacher Education, 16*, 211–236. doi:10.1007/s10857-012-9233-0

Desimone, L. M., & Le Floch, K. C. (2004). Are we asking the right questions? Using cognitive interviews to improve surveys in education research. *Educational Evaluation and Policy Analysis, 26*(1), 1–22.

Ferrara, S. (2007). Our field needs a framework to guide development of validity research agendas and identification of validity research questions and threats to validity. *Measurement: Interdisciplinary Research and Perspectives, 5*(2–3), 156–164. doi:10.1080/15366360701487500

Field, A. (2018). *Discovering statistics using IBM SPSS Statistics* (5th ed.). Thousand Oaks, CA: Sage.

Gall, M., Gall, J., & Borg, W. (2007). *Educational research: An introduction* (8th ed.). Boston: Pearson.

Gleason, J., Livers, S., & Zelkowski, J. (2017). Mathematics Classroom Observation Protocol for Practices (MCOP2): A validation study. *Investigations in Mathematics Learning, 9*(3), 111–129.

Hatch, A. (2002). *Doing qualitative research in education settings.* Albany, NY: State University of New York Press.

IBM Corp. (2016). *IBM SPSS statistics* [Computer software]. Armonk, NY: IBM Corp.

Jackson, K., Garrison, A., Wilson, J., Gibbons, L., & Shahan, E. (2013). Exploring relationships between setting up complex tasks and opportunities to learn in concluding whole-class discussions in middle grades mathematics instruction. *Journal for Research in Mathematics Education, 44*(4), 646–682.

James, L., Demaree, R., & Wolf, G. (1993). R_{wg}: An assessment of within-group interrater agreement. *Journal of Applied Psychology, 78*, 306–309.

Kane, M. T. (2012). Validating score interpretations and uses. *Language Testing, 29*(1), 3–17. doi:10.1177/0265532211417210

Koestler, C., Felton, M. D., Bieda, K. N., & Otten, S. (2013). *Connecting the NCTM process standards and the CCSSM practices.* Reston, VA: National Council of Teachers of Mathematics.

Krathwohl, D. R. (2002). A revision of Bloom's taxonomy: An overview. *Theory into Practice, 41*(4), 212–218.

Kruse, L., Schlosser, M., & Bostic, J. (2017). Fueling teachers' interest in learning about the standards for mathematical practice. *Ohio Journal for School Mathematics, 77*, 34–43.

Lance, C. E., Butts, M. M., & Michels, L. C. (2006). The sources of four commonly reported cutoff criteria. *Organizational Research Methods, 9*(2), 202–220. doi:10.1177/1094428105284919

LeCompte, M. D., & Preissle, J. (1993). *Ethnography and qualitative design in educational research* (2nd ed.). New York: Academic Press.

Leonard, J., Brooks, W., Barnes-Johnson, J., & Berry, R. Q., III. (2010). The nuances and complexities of teaching mathematics for cultural relevance and social justice. *Journal of Teacher Education, 61*(3), 261–270.

Matney, G. (2014). Deepening teachers' understandings of mathematical and pedagogical connectedness. *Journal of Mathematics Education Leadership, 15*(1), 50–59.

National Council of Teachers of Mathematics. (2014). *Principles to action: Ensuring mathematical success for all.* Reston, VA: Author.

Saderholm, J., Ronau, R., Brown, E. T., & Collins, G. (2010). Validation of the diagnostic teacher assessment of mathematics and science (DTAMS) instrument. *School Science and Mathematics, 110*(4), 180–192. doi:10.1111/j.1949-8594. 2010.00021.x

Appendix 8.A
Phases and Dual Knowledge Dimensions

The SMP-KA has four phases of knowledge assessment developed to engage the responder in five different dual dimensions of knowledge and cognitive processes (Anderson & Krathwohl, 2001).

Phase 1: Remembering factual knowledge of the SMPs (RFK)
Phase 2: Understanding conceptual qualities of the SMPs (UCK)
Phase 3: Creating scenarios of student engagement with factual knowledge *and* conceptual knowledge of the SMPs (CFK/CCK)
Phase 4: Analyzing scenarios of student engagement in the SMPs for conceptual knowledge (ACK)

Examples of Phases

Phase 1: Remembering Factual Knowledge

In this phase the assessment asks the respondents to list the titles of the eight practices. The list does not have to be word for word or in the common order found in the CCSSM (CCSSI, 2010).

> Example Response (**SMP1**): "Students should makes sense of each math problem they do and never give up trying to solve it."

Phase 2: Understanding Conceptual Knowledge

In this phase the respondents are given the eight SMP titles and asked to list qualities exemplifying each standard that are representative of mathematically proficient student behaviors for that standard.

> Example Response (**SMP1**): "Students try to figure out for themselves the meaning of a problem. Students look for entry points to a problems solution. Students analyze givens, constraints, relationships, and goals."

Phase 3: Creating Factual and Conceptual Knowledge

In this phase the assessment asks the respondents to give an example scenario of when a students is exhibiting engagement in the SMP. The respondents are instructed to describe the mathematical situation (problem) and explain how the student is exhibiting each SMP.

> Example Response for Decontextualizing in **SMP 2**: "The students were given the following problem: 'John and Mark brought oranges to school to share with the class. John was not sure how many he had but when Mark added his 16 oranges to John's every student in the class had exactly 1 orange of their own. If there are 35 students in the class how many oranges did John bring?' In considering this problem the students notice that they can make this problem into the equation X + 16 = 35. The students use the properties of algebra to solve for X noting they can take equal amounts away from both sides (take away 16) to get an equivalent equation revealing what John's amount should be. X = 19."

Phase 4: Analyzing Conceptual Knowledge

In this phase respondents are given a scenario in which students are engaging in mathematics problem solving. The scenario may have multiple SMP references but it has an overarching theme throughout. Respondents are asked to identify which SMP is themed by the scenario.

The following example is one of the pilot scenarios that was not chosen for the instrument. In order to preserve the integrity of the SMP-KA the scenarios in the instrument are not made public.

Example Scenario for **SMP 5**: Peaches Today . . . Peaches Tomorrow . . .

Students were given the problem: "A little monkey had 60 peaches. On the first day he decided to keep $\frac{3}{4}$ of his peaches. He gave the rest away. Then he ate one. On the second day he decided to keep $\frac{7}{11}$ of his peaches. He gave the rest away. Then he ate one. On the third day, he decided to keep $\frac{5}{9}$. of his peaches. He gave the rest away. Then he ate one. On the fourth day, he decided to keep $\frac{2}{7}$ of his peaches. He gave the rest away. Then he ate one. On the fifth day he decided to keep $\frac{2}{3}$ of his peaches. He gave the rest away. Then he ate one. How many did he have left at the end?" Students had access to snap cubes, fraction tiles, fraction circles, and counters. One student had worked on a similar fractions problem before and used counters. She went to grab the counters and try the strategy that worked before. As she was laying out the 60 counters she was realizing that it was taking a long time to organize that many counters. She decided that counters weren't going to be the most efficient strategy. She considered the other tools available and thought about fraction tiles or fraction circles, but the denominators in the problem do not match with the denominators on the tiles or circles. The tiles or circles do not

have $\frac{1}{7}$, $\frac{1}{9}$, and $\frac{1}{11}$ unit fractions. She then decided the best method was going to be paper and pencil. She began by drawing number lines to find the amount of peaches the monkey had at the end of every day.

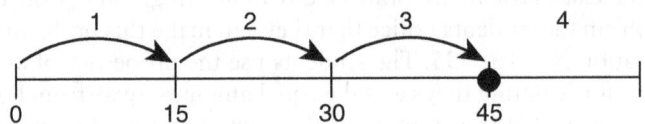

In order to find $\frac{3}{4}$ of 60, she first found $\frac{1}{4}$ of 60 by dividing 60 by 4 to get 15. Using her paper and pencil, she created a number line with 4 intervals of 15. She saw that 45 was $\frac{3}{4}$ of 60. On the first day, the monkey kept 45 peaches and ate 1. He ended the first day with 44 peaches.

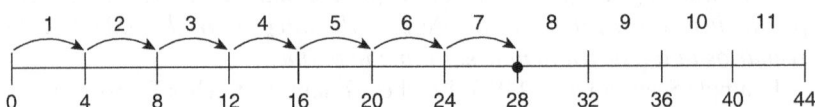

In order to find $\frac{7}{11}$ of 44, she first found $\frac{1}{11}$ of 44 by dividing 44 by 11 to get 4. She created a number line with 11 intervals of 4. She saw that 28 was $\frac{7}{11}$ of 44. On the second day, the monkey kept 28 peaches and ate 1. The monkey ended the second day with 27 peaches. The student continued to use the number line strategy for each day to determine that monkey had 1 peach at the end of the fifth day.

Appendix 8.B

Excerpts From Typological Analysis on the Five Knowledge Types

Knowledge Type	Expert Panel Data
Remember Factual Knowledge	"Most of my colleagues cannot even name, umm the SMPs. They have never read them and have no idea they are part of our standards for teaching math. I think if they even just knew that they would be better off because then they could think about whether or not they even asked their students to construct a viable argument or consider structure." "I, actually ashamed to say this, but I had never even read the Standards for Mathematical Practice prior to taking the PD on it and that was 3 years after state adoption. Eeek! I'm definitely not alone either. Teachers need to know about the SMPs. The titles are a great place to start . . ." "I think teachers need to have deep knowledge of the SMPs but most do not even know the names. When I ask them what SMP they want their students to engage in they often fumble around looking for the document that I sent them. I mean, they send me fully developed lesson plans but they haven't even considered the SMPs because they don't know the names much less what would be involved in instruction."
Understanding Conceptual Knowledge	"Teachers need to understand what it means to think about mathematics. The SMPs are about the way we engage in thinking about mathematics, the way we make sense of and create new mathematics. Understanding that is essential for teachers." "Because we have had PD I find the SMPs less intimidating. I can now, you know, classify and state what kinds of things students do when they do SMPs. So that's important for all teachers because like, what do the SMPs mean students should be doing? To know that is extremely helpful in every area, you know like, planning, teaching, and assessment." "Teachers should know what is in the SMPs; what exemplifies each one for students. I think they should be able to explain how students engage each practice. And be able to distinguish which students are engaging in it or not (comparing)."

(Continued)

Knowledge Type	Expert Panel Data
Analyzing Conceptual Knowledge	"What happens during instruction is a lot. As we are teaching there are many things to attend to with management, the content, and the practices. To do that job we [teachers] have to be able to quickly note what kinds of things students are doing that are attributes of the SMP they want to promote. It's so essential for us to think about this before, like we do in Lesson Study, because if we do not think about it before we don't notice that it's happening when it does."
	"So I think teachers should really be able to connect the pedagogical elements of teaching to student's mathematical thinking and ask themselves, 'How can I encourage them to engage in this SMP?' Or like, let's say that they watch a video or see another teacher teaching, [pause] if the teacher can point out how what is happening with the teaching and what the students are doing as evidence of an SMP then that would tell me that they 'know' the SMP."
Creating Factual and Conceptual Knowledge	"Essentially teachers create. We create plans to teach an idea that we hope students want to learn about and the SMPs are the same. We want to create the plan in such a way that students are doing these SMPs. When I didn't know about the SMPs it was the farthest thing from my mind to plan for them. But now after our Lesson Studies and everything we have done I know how to plan to help students do the SMPs."
	"I really think it would help all teachers to make sense of the SMPs and really think through what they look like in practice. That starts with their own engagement in mathematics but it can't just stay there. They [teachers] need to understand what it means to design lessons in which students are given rich mathematics tasks that draw out ways of encountering mathematics, you know, so that the students are doing the SMPs. Knowing how to visualize SMPs through the design of lessons is a must in my opinion because if we [teachers] can do that it would improve all students' math learning."
	"Teachers need to know how to create learning spaces in which students are not only thinking about mathematics content, but thinking about it in such a way that honors the SMPs. I think these practices are pivotal to student learning about any content. So what I mean is, teachers who have a good knowledge of the SMPs should be able to visualize and describe the kind of happenings during instruction that exemplify the SMPs."
Counter Evidence	"So like when an expert teacher knows how to really engage students in becoming mathematical thinkers I wonder whether or not they can do that even if they have no idea what the SMP titles are. It seems like they could. So while knowing the titles could be beneficial I'm not sure it's necessary."

9 Reflecting on the Past and Looking Ahead at Opportunities in Quantitative Measurement of K-12 Students' Content Knowledge

Jonathan D. Bostic, Erin E. Krupa, Michele B. Carney, and Jeffrey C. Shih

Introduction

Mathematics education research has explored K-12 students' knowledge for more than 50 years in a variety of ways. The focus of this chapter aims to provide readers with a sense of measurement of students' content knowledge outcomes: past, present, and future. It also describes challenges as well as opportunities that lie ahead for developers of instruments and measures assessing students' outcomes and those using quantitative tools. Chapters in this book reify some of these challenges and opportunities by describing various stages and facets of validation arguments. For the purposes of this chapter, we frame quantitative measures as those that produce a numerical value that can be operated on using analytical and descriptive statistics (Creswell, 2012). A key facet of outcomes from a measure is validity evidence and the associated validation argument, which is a central theme woven throughout this chapter. Of particular importance when measuring student knowledge, skills, and ability is providing sufficient evidence that a test or instrument can be interpreted and used in the manner described by the test developer. The chapter begins with a brief overview of measurement related to K-12 students' outcomes. Next, a discussion of trends ensues and finally, we highlight opportunities for further measure refinement and development. These discussions of trends and opportunities are actionable items, which mathematics educators must consider so that the future of measure refinement and development does not look like the past.

Two Lenses on Measurement of Students' Content Knowledge

To better understand what has been done, we purposefully chose to conduct two reviews of literature. The first search is meant to be broad, drawing across manuscripts using quantitative measures from one high-impact journal. The second search is meant to be narrow, drawing across research within one domain of mathematics education research: early childhood research. While other domains (elementary, secondary, problem solving, etc.) were viable, research with quantitative measures in early childhood mathematics is an area often explored by diverse scholars (e.g., educational, cognitive, and developmental psychologists, psychometricians and evaluators, and mathematics education researchers). Moreover, we had access to a reputable repository of peer-reviewed journal articles (https://dreme.stanford.edu/). In this section we present methods and results from these two different searches. Through these searches, we sought to examine how mathematics education researchers (including psychologists, psychometricians, and others) have developed and used content-focused instruments (e.g., measures), paying special attention to discussions of validity of the measures' outcomes.

Journal of Research in Mathematics Education: *Historical Analysis*

A literature search encompassing all peer-reviewed books, journals, and proceedings was beyond the scope of this project; therefore, the authors agreed to focus on a single high-profile mathematics education journal with a deep history among a wide variety of authors. Ultimately, the literature search was limited to articles published in the *Journal of Research in Mathematics Education* (JRME) between 1970 and 2017. JRME has been credited as the second most cited journal among mathematics education journals, having a very high impact factor, and the journal most familiar to mathematics education scholars (Williams & Leatham, 2017). The primary rationale for these dates was to explore validity and assessment in mathematics education literature through a historical lens, potentially uncovering changes over time within mathematics education research. The focus on published articles in JRME is that "JRME is the premier research journal in mathematics education and is devoted to the interests of teachers and researchers at all levels—preschool through college" (www.nctm.org/Publications/Journal-for-Research-in-Mathematics-Education/About-the-Journal-for-Research-in-Mathematics-Education/). Its wide scope and top-tier status in the mathematics education community (see Williams & Leatham, 2017) make it appropriate for a representative sampling of articles that included quantitative measures of students' content knowledge. These choices parallel a similar search by Hill and Shih (2009), conducted to examine validity with regard to quantitative measures used in JRME articles published between 1997 and 2006. Hill and Shih's sample space included articles focused on participants beyond students (e.g., teachers). Based upon Hill and Shih's prior work, we hypothesized that the current search would result in few articles discussing validity in ways that align with modern ways.

The search was conducted by one author and a graduate student. Each manuscript within JRME issues published between 1970 and 2017 was examined to determine whether a study used a quantitative measure of K-12 students' knowledge. Creswell (2012) describes quantitative measures as those that require descriptive or analytical statistics to answer a research question. In total, 2,175 articles were examined. Articles were retained or discarded based upon Creswell's criterion for quantitative research. Articles that included discussions of the validation process for a particular measure, validity evidence related to a particular instrument within the context of a particular students, and any suggestion of possible validity evidence were retained. Next, those that were retained were further examined with regard to types and approaches to validity and validation. Specifically, we sought evidence related to the five sources of validity: test content, response processes, relations to other variables, internal structure, and consequences from testing (American Educational Research Association, American Psychological Association, & National Council on Measurement in Education, 2014). In total, 97 articles met the criteria and were analyzed (4.5% of the total articles). Each article was reviewed for a discussion of instrument development of content measures with students and their use. Inductive analysis (Hatch, 2002) of those articles was conducted to generate themes. Theme generation included multiple reviews of the articles, seeking evidence and potential counterevidence for themes, and leading to a few central ideas that described the findings.

DREME Network: Analyzing a Single Domain's Recent History

We intentionally selected one domain to examine further and complement the JRME analysis. While there were many domains to consider, early childhood research using quantitative measures is conducted by many including mathematics educators; cognitive, educational, and developmental psychologists; and psychometricians. We recognize that inferences from this analysis may not be true for other domains (secondary mathematics, problem solving, etc.); however, it does provide a relatively reasonable window into present trends in quantitative research. In seeking to create a sample space, our team elected to delimit materials to the efforts of the Development and Research in Early Math Education (DREME) network based at Stanford University. Its mission is to "develop innovative tools that address high-priority early math topics and inform and motivate other researchers, educators, policymakers, and the public" (https://dreme.stanford.edu/mission). The list of publications provided on the network offers a broad scope of current research in early childhood mathematics, including a wide variety of quantitative journals such as *Cognition and Instruction, Journal of Cognition and Development, Developmental Cognitive Neuroscience, Journal of School Psychology,* and *Developmental Psychology.* Many of the peer-reviewed manuscripts found on the site were not in traditional mathematics education journals.

We purposefully delimited our search to peer-reviewed articles in the DREME network to parallel our review of JRME articles. Thus, it was advantageous to take a different perspective on what measures were used and by whom.

The DREME network's resources consisted of books, book chapters, conference presentations, and peer-reviewed journal articles published between 2014 and 2018. Peer-reviewed articles were published after the *Standards for Educational and Psychological Testing* (*Standards*; American Educational Research Association, American Psychological Association, & National Council on Measurement in Education, 2014); hence they had a greater likelihood for including attention to validity. Additionally, they were found in a variety of peer-reviewed journals that include a diverse audience of authors and readers, such as psychology and measurement experts. Our hypothesis was that this complementary analysis might indicate how others conducting scholarship in mathematics education may attend to validity within measurement of students' content knowledge. In sum, 24 peer-reviewed journal articles inclusive of measurement of K-12 students' mathematics knowledge within early childhood became the sample space for the search. Inductive analysis (Hatch, 2002) was conducted on the journal articles by two authors, differing from the pair examining the JRME materials (i.e., the graduate student from the earlier pairing was not part of the DREME review). To clarify, we use "inclusive" in the preceding sentence because some studies described professional development and other interventions on teachers' instructional practices and the outcome variable under investigation was students' outcomes.

Following the analyses, the three individuals came together to discuss whether themes might be consistent across the two searches. Surprisingly, two major themes were identical across the two analyses. One theme was that validity evidence, when provided, was mainly discussed in the context of test content and internal structure, which typically was reported as a measure of internal consistency. The lack of discussion about other sources of validity evidence is intentional because information related to those sources was not present in the published articles. A second theme was that validity, when discussed by authors, generally described the measure rather than the scores. These themes will be discussed in more detail in the following sections.

Validity Evidence: Test Content and Internal Structure

The *Standards* (American Educational Research Association et al., 2014, 1999) indicate that validity evidence should be provided for quantitative instruments. For example, discussions of some or all of the sources of evidence are warranted (e.g., content, response processes, relations to other variables, internal structure, and consequences from testing).

Greater evidence within and across domains supports stronger—or more robust—validity arguments. Unfortunately, few articles included discussions of validity evidence. This is significant because a lack of validity evidence and robust validation arguments calls into question the generalizability of research findings. Put simply, if it is unclear how researchers "know" what they know based on the quality of their research instruments, then how can other scholars have confidence in a study's outcomes?

Between 1970 and 2017, seven JRME articles out of 97 (7.2%) mentioned validity evidence within the manuscript. Three articles discussed both content and internal structure evidence, three discussed content evidence only, and one article was a factor analysis study focusing on a single measure (i.e., internal structure). Between 2014 and 2018, seven articles out of 24 (29.1%) from the DREME network indicated some form of validity evidence. One article described multiple forms of validity evidence while the remaining six articles described evidence from a single source of validity (i.e., content, relations to other variables, or internal structure). When reliability was reported, it was typically stated as a form of internal consistency. Regarding reliability, 24 of 97 JRME articles (24.7%) stated the reliabilities of the measures used in the study. Eleven of 25 articles (45.8%) from the DREME network report a reliability statistic. Cronbach's alpha and KR-20 values were most commonly found in manuscripts. Taking a broad perspective across both samples, it is evident that most manuscripts examining students' outcomes do not report validity evidence associated with inferences and scores associated with student-related measures. While reliability was found more often across our samples, still it was not communicated more times than it was.

Provided below are a few brief examples of JRME articles lacking adequate discussion of validity evidence. Post and Brennan (1976) constructed a problem-solving test from various sources but did not describe any review by content experts. They calculated a KR-20 score as a form of reliability, which is a form of validity evidence. However, using classical testing theory approaches means that the reliability score is connected to the study's sample, not the population of interest (Crocker & Algina, 2006). Because the study was of a particular sample and not generalizable to a population of interest, this internal structure evidence may disassociate the outcomes and measure in a way that the reported reliability was tied to the measure's scores with a particular sample. Shumway and colleagues (1981) explored students' content knowledge as an effect of calculator use. A test of basic facts was created and its reliability was reported. A mathematics achievement test was selected from the Stanford Achievement Test series (1972–1973). The tests were not described, and again, internal consistency was discussed as having been reported elsewhere. Selke, Behr, and Voelker (1991) explored multiplication for middle school students and created a pretest and posttest, which were different but similar in structure. Here again, tests were discussed but their related content validity evidence was not presented in the article.

The *Standards* (American Educational Research Association et al., 2014, 1999) advocate that instrument users and developers are responsible for providing evidence to function as part of an argument justifying that the study's outcomes are logically drawn. There was limited evidence that some researchers valued collecting content and internal consistency validity evidence. For example, Roberge and Flexer (1983) sought to examine mathematics achievement of sixth, seventh, and eighth grade students. They used three tests, one developed by the authors and two developed by other researchers. Regarding the one developed by the authors, they presented content validity evidence with one sentence connecting it to a framework. Content validity evidence, when located within the articles, was usually related to researcher-created measures. Internal structure was usually reported for a sample for these measures. Again, Roberge and Flexer (1983) state: "they [authors] presented factor analytic evidence of the FORT's construct validity. They reported test-retest reliability coefficients . . . they also reported internal consistency reliability coefficients for samples" (p. 347). When researchers used standardized measures, it was typical to indicate that validity and reliability evidence was reported elsewhere. For example, Roberge and Flexer (1983) stated that the Group Embedded Figures Test (GEFT) was used. They noted that others, including instrument developers, "described the GEFT as a valid and reliable alternative [measure]" (p. 346). Roberge and Flexer reported reliability coefficients for these measures and surprisingly commented that the reliabilities were for a different sample from theirs, and that validation was not done for their sample—without reporting validity or reliability evidence for their own sample/population. In sum, discussions of validity evidence were rare but prominent enough to substantiate a theme that content and internal consistency were ways that researchers were framing validity discussions related to quantitative measures of content knowledge.

A few brief examples from the DREME review are shared to connect with the theme that content evidence and reliability were rarely discussed, if at all. Chan and Mazzocco (2017) designed the Attention to Number (AtN) task, a picture-matching task used to assess attention to number relative to other features. To measure number knowledge up to five, they used modified items from the Test of Early Mathematics Ability–Third Edition. There was no validity evidence provided beyond context of the test content, and reliability was not reported (Chan & Mazzocco, 2017). Watts et al. (2017) used the Research-Based Early Math Assessment (REMA) to measure mathematics achievement across a wide range of concepts and procedures. The measure was "extensively validated . . . through repeated testing in three different samples, which produced an overall reliability of .94" (Watts et al., 2017, p. 103). The justification of the use of the multiple-choice TEAM 3–5 test, used to assess mathematics achievement of third through fifth graders, was its internal reliability and high correlations with state achievement tests. It was unclear from the article whether the correlations were related to similar content or differing outcomes. While having a connection to other measures is important as a source of relations to other variables, the evidence provided did not necessarily convey confidence using the REMA in this study within the validity argument. Connecting back to the JRME review, reliability was typically presented as a form of internal consistency, not using the framing described in the *Standards* (American Educational Research Association et al., 2014). It was common to read statements like "the measure was validated" in many manuscripts, which adheres to an outdated view of validity and validation. Such statements contradict modern notions of validity, validation frameworks, and validity arguments where current thinking reflects a more rigorous but fluid view of validation. Validation is not something carried out at one instance and never reviewed again (Kane, 2016). This modern notion of validation described in the *Standards* (American Educational Research Association et al., 2014) encourages gathering a collection of evidence that expands but is never a done deal for all cases—every new application of a measure or change in how a measure is used requires a reframing and consideration of the available evidence for score interpretations.

Validity of Tests Versus Validity of Score Interpretations From Tests

It was common for articles to discuss validity in terms of the test rather than in terms of the scores' interpretations, including ways that reliability was reported. We present support for this theme from the JRME and early childhood analysis. Every article but one in both searches associated the validity evidence as part of the tests' properties and not the outcomes of the measures.

JRME

Szetela and Super (1987) used seven quantitative content measures in their study of calculator use and problem-solving outcomes. Two measures had been developed by others whereas five were internally developed. For the two measures developed by others, they report the reliability coefficient (Cronbach's alpha) for the test as 0.88 and 0.92. Similarly, Cronbach's alpha values were reported for the internally developed tests: 0.75, 0.72, 0.78, and 0.77. The final internally developed measure did not have a reported reliability value associated with it. There were instances where the focus was on the test rather than the outcomes from the test. Two relatively recent examples are shared here, and both use standardized measures (i.e., assessments developed by others).

Sample McMeeking, Orsi, and Cobb (2012) examined the effects of a professional development program on students' outcomes. They used measures from the Colorado Student Assessment Program to measure students' mathematics achievement. They described it as a standards-based assessment with constructed-response and multiple-choice items. The authors stated that

> Extensive psychometric testing has been completed to ensure the validity and reliability of each test and the process by which it is graded . . . also required . . . content and construct validity of the Colorado Student Assessment Program, which included reviewing the curriculum, [and] examining each test item for content and bias.
>
> (Sample McMeeking et al., 2012, p. 167)

Here, the authors describe validity of the Colorado Student Assessment Program and mention that content evidence was explored, but there was no further discussion, including mention of other peer-reviewed work or white papers that connect students' outcomes to the test. Like Gavin, Casa, Adelson, and Firmender (2013), discussed above, the authors again frame validity and reliability as a property of the test. In all of these instances, the reliability coefficients are devoid of a contextualization; thus, it is difficult to discern how these scores reflect the validity of the respondents' outcomes.

Early Childhood Research Within DREME

Tosto et al. (2017) used data from a longitudinal study to investigate the relationship between number sense and mathematical ability/performance across time. They used 11 computerized tests that measured estimation, mathematical performance, general cognitive ability, language ability, and reading ability. The authors stated that

> Prior to the main data collection, the tasks were piloted and tested for reliability and suitability for Web administration using samples of 16-years-old singleton and twin students. All tests proved to be suitable for Web administration . . . and showed good internal consistency and test-retest reliability.
>
> (p. 1928)

This language was common; its focus was on the administration of the test rather than the outcomes of the test. A second example, Gunderson, Park, Maloney, Beilock, and Levine (2018), used multiple measures to assess students' motivational frameworks, math anxiety, math achievement, and reading achievement. The 10 reliability coefficients (i.e., Cronbach's alpha) were reported, ranging from 0.60 to 0.94. As with the

JRME review, these reliability coefficients were presented without contextualization, making it difficult to discern how these scores reflect the validity of the respondents' outcomes.

Rather than continue to share more evidence, we felt there is value in highlighting a counterexample for readers to better understand an effective way to convey evidence about a measure's outcomes. This measure, the Concepts and Applied Skills Assessment (CASA), was used within a study of second grade students. Its outcomes describe students' mathematics knowledge and is appropriate for K-5 students (Connor et al., 2018). The discussion of the CASA is unique in that the outcomes' implications were discussed and flowed from the developers' intentions.

> The 93 items in the CASA are divided into sections by mathematical skill areas. These skills include: numeration, measurement and data analysis, fractions, geometry, and word problems. Each section includes items that increase in difficulty to determine students' proficiency and instructional levels. Subscales and subscores are also available for each section. Reliability for the total score with this sample was excellent (alpha = 0.92). Reports of the CASA were shared with teachers . . . that immediately followed administration, with four administrations over the school year. Reports included results for each student in table form, focusing on skills mastered and not mastered . . . showing gains in scores over the school year. For some students, particularly those who were not making expected progress, the teachers and research partners would review the entire test. The research partners and teachers together decided how to group students and which aspects of mathematics to focus on during stations. . . . The CASA was specifically designed to facilitate this kind of instructional decision-making.
>
> (Connor et al., 2018, p. 103)

The first four sentences describe the measure itself and what it does, elements that should be found in a purpose statement (see Kane, 2012 for purpose statement). Those sentences also tell readers about how to use the measure, something not found in any other manuscript reviewed in either sample space. Next, the authors correctly frame reliability for their scores in the sample, rather than as a test property. Finally, the discussion of the measures' output and consequences of testing evidence are connected to its purpose, as a formative assessment. The readers learn quite a bit about the measure's outcomes in eight sentences. This counterexample has language that is clear and adheres to best practices as advocated by the *Standards* (American Educational Research Association et al., 2014). Thus, it might serve readers as a way to convey appropriate ideas about a measure, its outcomes, and connect to the *Standards* (American Educational Research Association et al., 2014).

Moving Forward From Results of These Reviews

The aim of these analyses is to portray how validity evidence, related to quantitative measures of K-12 students content knowledge, have been portrayed in one domain and a respected journal. A limitation is that it is not exhaustive across all venues; but, it provides readers with a view into what has been acceptable within a prominent journal. We do not intend to criticize the current or previous editorial staff or published authors. Instead, we aim to share what has been common practice. JRME as a journal is not focused on measure development. It is an appropriate reflection of how much thinking about validity is found in the most influential research on math education, which might or might not be using quantitative measures. JRME is holding scholars to higher standards; however, validity arguments and connections to the *Standards* (American Educational Research Association et al., 2014, 1999) are rare. A second limitation is that this team of authors cannot necessarily decipher whether the authors of articles in the sample space gathered validity evidence for their measures or whether editorial decision making led to not including validity evidence related to the measures. A secondary literature search in mathematics education and assessment journals related to each measure might be taken up for a particular measure to discern this. In any case, these results provide one viewpoint for consideration related to validity aspects related to quantitative measure of students' mathematics knowledge.

As indicated by the first theme, if validity is discussed within a manuscript then most mention researchers discussed content and/or reliability within the instruments section of their article, if presented at all. This was true in both the JRME and DREME review. At this time, it is important to foreground these results within the context of *Standards* (American Educational Research Association et al., 2014, 1999) in which both have similar sources of validity evidence: content, response processes, relationship to other variables, and internal structure. Consequences from testing is mentioned in the 1999 version; however, it garnered far more attention in the 2014 version of the *Standards*. One article does not necessarily need to present evidence for all sources within the context of a validity argument; yet if the selection of articles reviewed is representative of some of the most impactful research from the past 40 years as well as a significant domain, then an implication is that we are not, as a field, close to collectively investigating the breadth and depths of evidence that would make up a strong evidentiary basis for the measures used within quantitative studies of students' outcomes.

As indicated by the second theme, there are still many scholars drawing upon an outdated notion of validity being tied to a measure and not its outcomes/scores. However, there is evidence that some authors (e.g., Connor et al., 2018), and others in this book and *Assessment in Mathematics Education Contexts: Theoretical Frameworks and New Directions* (Bostic, Krupa, & Shih, 2019) are drawing on modern notions of validity evidence and arguments. Thus, some measure developers and users are paying attention to this important detail. An implication tied to this second theme is that further work is needed to support everyone working in the mathematics education space to connect with language found in the *Standards* (American Educational Research Association et al., 2014) as well as validation argument frameworks.

Validity may not be at the forefront of scholars' minds during instrument development. Yet as evidenced by this book, there is a growing number of scholars who are keeping validity in their minds during measure development and that has potential to shape the future of quantitative research in mathematics education. These results present an opportunity for mathematics educators interested in assessment and early childhood to explore further. Related to this opportunity, we highlight some recent trends in measurement of student knowledge that might be taken up by scholars at the nexus and periphery of mathematics education research.

Trends in Assessment of Student Knowledge

The purpose of this section is to describe two recent trends in measurement of K-12 student knowledge and to situate these trends within a contemporary approach to validity and validation. These trends come from current research in the education community: the instructional relevance of assessments and assessing student cognition. By contemporary approach to validity and validation, we are primarily referring to (a) thinking about validity as referring to proposed interpretation and use for an assessment as opposed to validity being a quality of an assessment, and (b) validation as a process of articulating an argument in support of the proposed interpretation and use (see American Educational Research Association et al., 2014). This argument involves the explicit articulation of assumptions underlying the proposed interpretation(s) and use(s) along with presenting validity evidence in support of those assumptions (see Kane, 2001, 2006, 2012, 2016). First, we briefly describe the trends from the perspective of the proposed interpretation and use. Then, we describe potential assumptions that would need to be investigated based on the interpretation and use. Lastly, we describe potential sources of validity evidence if applicable.

Instructional Relevance

Formative assessment has become a prominent trend in education (Black & Wiliam, 2009; Fennell, Kobett, & Wray, 2017). Formative assessment practices assume that assessment results can be interpreted as having instructional relevance and can be used to information instructional practice (Black & Wiliam, 2009; Fennell et al., 2017). This connects to more current conceptions of validity referring to score interpretation and use—as opposed to how well a test purports to measure a construct. However, there are numerous assumptions inherent in the claim that assessment results provide relevant information to inform instructional practice. There assumptions are often underinvestigated. Pellegrino, DiBello, and Goldman (2016) state:

> Despite the significant increase in efforts to frame the focus of discussions of assessment in terms of domain-based theories and models of cognition and learning, and an increasing attention to the use of assessment in the classroom to support various teaching and learning functions, there has been a paucity of discussion about the meaning of validity for assessments intended to function close to instruction.
>
> (p. 60)

Assessments of student knowledge are typically assigned a quantitative score. These scores can be difficult for classroom teachers to use to meaningfully inform classroom instruction beyond perhaps a general sorting of students into categories of high, medium, and low levels of knowledge for a particular topic (e.g., for summative grading purposes). This general sorting does not assist teachers in better understanding what students do and do not know about a topic or how instruction could be potentially scaffolded for students in the different groups (i.e., formative purposes). Shepard (2018) states,

> To support student learning, quantitative continua must also be represented substantively, describing in words and with examples what it looks like to improve in an area of learning. For formative purposes, in fact, qualitative insights are more important than scores.
>
> (p. 1)

Therefore, our identified assumptions build upon this initial claim that in order for measures of student knowledge to have a high-quality impact on instruction, a resulting quantitative score needs to be translated into a qualitative interpretation (e.g., a mapping of mathematics skills or ideas). The underlying assumptions related to this claim are that the resulting qualitative interpretation is (1) accurate, (2) useful for classroom teachers to inform practice, and (3) focuses classroom instruction on important mathematical ideas and concepts. We do not have the space here to fully investigate each of these assumptions; hence we provide a brief example of why it is important to investigate these assumptions using the accuracy assumption as an example.

Many assessment developers now provide qualitative interpretations of student performance on assessment (e.g., NorthWest Evaluation Association's Measure of Academic Progress (MAP) assessment or Renaissance Learning's Star Math); but what evidence should they provide related to the accuracy of interpretations? Item anchoring—examining items with similar difficulty levels to identify common mathematical topics or concepts—is often used as a means of developing and justifying qualitative interpretations of quantitative continuum (Shepard, 2018). However, because item anchoring methods are typically conducted post hoc, it is difficult to isolate that a particular mathematical topic or concept is the primary source of the variance rather than other factors, such as the number relationships or problem context. For example, Carney, Smith, Hughes, Brendefur, and Crawford (2016) found that number relationships can significantly influence item difficulty on otherwise similarly structured proportional reasoning items, resulting in a wide variance of item difficulties for a particular topic. How this variance is translated into an accurate qualitative interpretation using an item anchoring approach needs to be well substantiated by supporting evidence. While item anchoring can provide some evidence in support of the accuracy of a qualitative continuum for interpretation, further evidence is likely needed to substantiate the accuracy of the interpretation—perhaps through response process evidence—that students grouped under a particular score can be interpreted as having common understandings of particular mathematics topics and ideas. Synthesizing across these ideas, researchers developing validity arguments must be flexible because assumptions within their research design may present unusual challenges. To that end, we advocate drawing upon validity argument frameworks (e.g., Kane, 2001; Pellegrino et al., 2016; Wilson, 2005) that fit a desired research program rather than choosing one at random. Additionally, it is necessary to gather appropriate evidence to support the interpretations about instructional relevance that in turn have potential to impact classroom practices and student learning.

Focus on Student Cognition

Historically, school mathematics has positioned student fluency with mathematics procedures and algorithms as a primary focus of instruction and learning (Kilpatrick, Swafford, & Findell, 2001). More recently, there has been a shift to mathematics instruction focusing on students fluently and flexibly using mathematics procedures as well as reasoning and making sense of mathematical situations more generally (Kilpatrick et al., 2001; National Council of Teachers of Mathematics, 2009, 2014, 2018). This increased focus on student conceptual understanding of mathematics has led to calls for assessments that focus on student cognition (National Research Council, 2001). As researchers conduct interventions designed to influence specific aspects of student conceptual understanding (often mathematics domain specific), they need instruments that are closely aligned to the aspect of cognition they are trying to influence. There are seldom assessments available to these specific domain aspects of cognition. Therefore, researchers need to develop these instruments as part of their research. But what assumptions are inherent in stating that test scores can be interpreted in relation to student cognition?

A key underlying assumption of interpreting scores related to students' cognition is that the student cognition construct has been operationalized in such a way that it can accurately represent the construct under investigation. The domain of fractions provides a useful illustration of the complexity of this assumption. When thinking of developing an instrument to assess fraction knowledge, one may decide to focus on students' ability to accurately perform fraction operations. This could be operationalized rather simply in a matrix that focuses on the (a) four operations, (b) the complexity of the denominator conversions required in the problems, and (c) the use of bare number problems and contextual problems. This conceptualization of fraction knowledge involves the development of a relatively simple test blueprint and is a common instrument operationalization of fraction knowledge (e.g., in a textbook manufacturer's assessments). This conceptualization focuses on students' ability to perform fraction operations and does not focus on student cognition related to fraction understanding.

On the other hand, a conceptualization that focuses on student cognition related to fraction understanding might focus on students' underlying conceptions related to fractions. Take for example the idea that students need to understand the meaning of the numerator and denominator in order to perform operations with conceptual understanding. In particular, they need to understand that the denominator denotes the quantity and size of the pieces in a unit of one, and that as the number of pieces increases, their size actually decreases. In addition, students need to understand that the numerator indicates the number of pieces of that particular fraction size. Students can perform operations with fractions without this conceptual understanding. There is strong evidence to suggest that by not conceptually understanding these ideas—and performing operations based on memorized procedures—students' abilities to correctly apply the procedures and use them effectively in new situations is greatly reduced (Kilpatrick et al., 2001). Therefore, it is important to assess conceptual understanding of student cognition when examining their understanding of fractions. Operationalizing ideas such as this into assessment maps and items is relatively complex. Assumptions related to the assessment of complex cognitive construct requires extensive evidence regarding how the domain is conceptualized and operationalized. While the assumption that an instrument is sufficiently assessing students' ability to perform fraction operations may be reasonably well-supported through a defined test blueprint, assumptions related to sufficiently assessing students' conceptual understanding related to fractions requires additional evidence. That evidence may describe (a) how the construct is theorized/conceptualized, (b) how that theorization/conceptualization is translated into an assessment blueprint with a focus on addressing potential issues of conceptual underrepresentation, and (c) that the items pinpoint aspects of cognition intended to focus on addressing potential issues of construct-irrelevant strategies being used. One might say that these assumptions need to be addressed in all testing situations. For assessment of complex constructs, they become particularly important and require more time and effort than may be typical (e.g., the gathering response process validity evidence to demonstrate students are not using construct-irrelevant strategies).

As presented in this section, there have recently been two trends related to quantitative measures of students' content knowledge. Historical trends and analyses provide leverage for predicting and suggesting what others might consider in the future as viable opportunities to pursue. Thus, the next section offers research ideas for scholars, industry leaders, and the next generation of researchers to consider as they work within this space of quantitative content measures of students' knowledge.

Vision for Future Measures of Students' Knowledge

As the format and constructs of current assessment models evolve, so should their associated validity arguments. However, Mislevy (2016) argues, through this evolution, validity arguments have largely remained the same and some forms of evidence are progressing faster than standard validation techniques in the field. In addition, the use of technology is changing the landscape of assessment and the scoring of assessments. Mislevy (2016) stated, "these developments (psychology and technology) expand the range of evidence and theory we must address in validation activities, and the argumentation through which we marshal them" (p. 265). Further, more complex assessments are being created and customized to individuals through computer adaptive testing and performance-based assessments. Mathematics educators (inclusive of all scholars working in the mathematics education scholarly space) may need to change their assumptions within their validation arguments because of a continually changing landscape that evolves as technology improves. This section explores some of the ways in which emerging technology and scoring systems, as well as newer forms of assessment, influence validation arguments and questions that arise from such arguments.

Education Assessments of the Future

Gorin (2014) argues educational assessments from the 20th century were limited by focusing "on relatively simple claims and often rely on a single piece of evidence" (p. 1) and that the field should move to evidentiary arguments that come "from multiple evidential sources to make the most valid and reliable claims about student learning" (p. 3). The future of educational assessments should focus multiple sources of evidence gathered for validity arguments to guide the interpretation and use of test scores. Traditional standardized assessments rely on a single score from a single source, whereas Gorin points out that alternative assessments for students with severe cognitive disabilities rely on multiple, and varied, sources of evidence drawn from behavioral observations, performance assessments, rating scales, portfolios, past performance, and background information. Here is an example of how traditional ways of thinking about validation may evolve because of different foci under investigation. It is possible that future student mathematics assessments will consist of evidence gathered from multiple sources, including those that harness the power of technology and those that encapsulate a performance-based or portfolio assessment. These could take the form of a system of assessments where evidence is gathered from multiple measures and used jointly to make decisions about student learning. Validation arguments from a single source of evidence are complex; imagine the challenges to validity arguments that might arise from an assessment system with multiple forms of evidence.

The nature of assessment is changing so rapidly that the line between curriculum and assessment is blurring, potentially because of the importance of instructional relevance and the increased mechanisms for measuring formative assessments. One of the benefits of technology that Shute (2011) notes is the ability to "embed assessments into the learning process" (p. 504). When assessments are embedded into the curriculum, careful attention needs to be paid to distinguishing the assessment and creating a coherent validity argument for the student learning that occurs. The assessment and learning foci have different purposes, and therefore the underlying assumptions will differ. However, both assumptions for assessment and learning need to be investigated. In addition, the claims that researchers want to make about student learning may change as a result of new technology and because assessments are being embedded into the teaching and learning of mathematics. This changes the data types that are gathered, which can result in scoring challenges and a need to analyze current conceptions of validity. Questions about the validity of scores from tests include: How can scores be translated into a qualitative interpretation to be used for instruction? How can we move from a single source of evidence to multiple sources at scale? How can the power of technology be harnessed to improve the interpretation of test scores and use?

Use of Technology

The use of technology warrants changes to validity arguments and to the sources of validity evidence necessary to support the validity argument. However, as Gorin points out, the "transitions from paper-pencil to computer administration of multiple-choice tests merely changed the delivery mode of the identical items, rather than a change to the nature of the tasks and the measured constructs" (2014, p. 16). Future assessments should harness the power of technology to provide new mechanisms for tasks and innovative response types, which would influence how constructs are measured and how items are scored.

As technology evolves, the landscape of assessments is changing due to the nature of test administration and content delivery via a computer, calculator, or online system. Validation of claims made about student learning from assessments administered using these systems should evolve as well. However, above all else, the most dramatic shift according to Gorin (2014) is "due to the way in which we collect evidence to support those claims" (p. 17). The two sections below further examine ways in which the use of technology is changing the types of validity evidence gathered and new forms scoring assessments.

Sources of Evidence

Technology has the opportunity to revolutionize the evidence that is collected through computer-based assessments and through curricula delivered electronically. In several studies on the influence of learning in a computer-based environment on students' understanding of algebra, researchers were able to collect the following data sources from the software system: scores for homework, quizzes, and final exams; the length of time students spent in the computer-based environment on content modules; and how much time students spent on homework, quizzes, and tests. Researchers were able to embed a Likert-item survey into the computer-based environment (Krupa, Webel, & McManus, 2015; Webel, Krupa, McManus, 2017). Similarly, Confrey and Toutkoushian (this volume) gather digitally scored diagnostic data on students' progress along learning trajectories and provide detailed score reports to students and teachers. The data collected in these studies provides automated scoring of items and evidence of how long students spend on assignments or in course modules, and gives students and teachers an automatic visual of individual student's progress with the assessed content material. These are additional data sources that can be gathered through the assessment administration, which could be used as evidence in a validation argument. To use such evidence, the underlying assumptions the evidence is being used to support would need to be clearly articulated. Ultimately, the new ways of collecting evidence of a skill also change the arguments that need to be made regarding validity.

Diagnostic assessment systems "must generate evidence of student processes including strategy use, processing speed, attentional control, response selection, and self-regulatory processes" (Gorin, 2014, p. 15). Generating new sources of evidence will change the ways in which researchers structure validity arguments. Some researchers are gathering new sources of evidence. Wise and Demars (2006) used response times as evidence of student motivation and engagement. They claim that rapid guessing or low engagement with an item can be determined by short response times, which influence score interpretations (Demars, 2007). Their validity argument was shaped by analyses of a test scored with an effort-moderated model, correlation between the course grades and students' number of rapid guesses, and subsequent validity coefficients.

Researchers are also using log-data collected in a computer-based environment as sources of evidence for validity claims. Gobert, Sao Pedro, Raziuddin, and Baker (2013) collect log-data from online microworlds to detect science inquiry skills on a performance assessment. Behrens, Mislevy, Bauer, Williamson, and Levy (2004) have used simulation-based assessments to measure networking proficiency, and log-data is collected to provide evidence of students' problem-solving abilities. Log-data include the sequences of keystrokes, mouse clicks, and scrolling made by the student, all of which can be time-stamped. These data are much more complex than standard assessment data and raise questions about conditional independence (Mislevy, Behrens, Dicerbo, and Levy, 2012), lack of theory for analyzing the data in these environments (Gobert, Sao Pedro, Baker, Toto, & Montalvo, 2012; Quellmalz, Timms, & Schneider, 2009), and the establishment of criteria to analyze patterns derived from log-data (Gorin, 2014).

Another novel form of data used in validity arguments is eye-tracking data. Studies utilizing eye-tracking data of students' problem solving have provided evidence connected to students' cognition and attention (Gorin, 2006, 2014; Ivie, Kupzyk, & Embretson, 2004). The *Standards* refer to eye-tracking data as a source of validity evidence, specifically evidence based on response processes: "documentation of other aspects of performance, like eye movements or response times, may also be relevant to some constructs" (American Educational Research Association et al., 2014, p. 15). Gathering evidence such as eye-tracking data and response times may require new scholarly collaborations with learning sciences, psychologists, and/or computer scientists, to name a few. Through these new collaborations, validity arguments will be current and move away from traditional approaches as well as the unfortunate history of merely collecting test content and reliability evidence for mathematics measures. Taken collectively, data collected from computer-based systems can provide evidence for validation arguments. However, these sources also raise some important questions for validation teams to consider: What research needs to be conducted to warrant the use of these sources of evidence? How can researchers score novel data sources that have different structures than traditional sources? How will validity arguments change as a result of new sources of evidence? What evidence is needed to support that eye-tracking data can influence a construct? How do we interpret the data to support that construct? What assumptions underlie that interpretation? What does construct-irrelevant variance mean for eye-tracking data?

Scoring Systems

While automated scoring systems have been in development in mathematics for many years (Bennett & Sebrechts, 1996; Sebrechts, Bennett, & Rock, 1991), new innovations continue to change the manner in which assessments are scored. Constructed-response items are becoming common among high-stakes assessments. The increased prevalence of constructed-response items is a result of improved technologies that allow for the automated scoring of the responses (Shermis & Burstein, 2003). As part of the release of the Common Core State Standards (CCSSI, 2010), the Race to the Top movement encouraged the assessment consortia to utilize automated scoring for both literacy and mathematics constructed-response items (Bennett, 2011). In 2011, a conference was convened to ensure experts from the mathematics education community were involved in the development and implementation of these new assessments, a group of mathematics curriculum and assessment developers, policymakers, and state and district mathematics specialists. Two notable recommendations from the conference include (1) to develop technology with the capacity to administer assessments equitably at scale, paying careful attention to how assessments are scored and (2) to assist the consortia in the creation of scoring categories, subscores on constructs, and tagging systems to ensure valid information is being reported to teachers, parents, and students (Krupa, 2011). Developing new technology to support the automated scoring of items could potentially standardize the scoring of assessments, however, as we see below, each item type has its own set of challenges when considering automated scoring.

Mathematics assessments have unique item types that have implications for scoring. Bennett, Steffen, Singley, Morley, and Jacquemin (1997) introduced the mathematical expression response type, where the student is asked to provide a mathematical expression as their answer. These types are difficult to score due to the unlimited number of ways of expressing equivalent relationships (Clauser, Kane, & Swanson, 2002). In addition, Bennett (2011) describes three additional task categories:

> Those calling for one or more instances from a potentially open-ended set of numeric, symbolic, or graphical responses that met a given set of conditions; those requiring the student to show the symbolic work leading to the final answer; and those asking for a short text explanation.

(pp. 2–3)

Bennett argues that the accuracy in scoring across these four types varies drastically, as does the ability for computer scoring, and he provides a list of scoring quality reports and key limitations in automated scoring across the four types. There is limited data on the scoring of short text responses and on items from an open-ended set of numeric, symbolic, or graphical responses, however mathematical expression responses and symbolic work leading to a final answer have scoring comparable to that of traditional assessments (human scoring and machine scoring, respectively) (Bennett, 2011).

Though automated scoring is becoming more prominent, there is still some contention around the accuracy and use of computerized scoring. Way, McClarty, Murphy, Keng, and Fuhrken (2011) argued that due to the "diversity of problems and responses that are conceivable . . . in mathematics suggests that the comprehensive use of automated scoring will not be feasible" (p. 24). Clauser et al. (2002) contend that the "use of automated scoring systems has the potential not only to enhance score validity but to introduce serious threats to validity" (p. 430). Several questions regarding scoring remain: How do we overcome the threats to validity that automated scoring presents? Due to limited data on scoring systems for certain types of mathematics items, how do researchers create sound validity arguments when using scoring systems? How do scoring systems transform the interpretive argument?

Summary

Mathematics researchers and assessment developers are at a junction where the capabilities of technology are progressing at a much faster rate than the theories of validation can keep up. It presents opportunities for collaboration across fields and to develop robust knowledge that can implicate scholarship in a way that has broad impact and strong intellectual merit. We presented emerging ideas on the future of student measures in mathematics education and raised questions about the use of technology and automated scoring of assessments for validity arguments. Researchers should consider how new forms of assessment change the way validation is approached.

Final Thoughts

The purpose of this chapter was to provide readers with a sense of measurement of students' outcomes: past, present, and future. This chapter may also disturb readers because it portrays what happens when validity evidence and validity arguments are lacking. Without attending to validity evidence and arguments, it is difficult to discern the degree to which implications and conclusions are logically drawn from the measures used. Therefore, we as a field of scholars working within the space of K-12 students' mathematics content knowledge are struggling to answer a major question because of a lack of attention on validity: How do we know what we want to know, to be appropriately drawn from the measures used in a study? The future of measure refinement and development should not look like its past. This chapter does not intend to be comprehensive, but it should invigorate readers interested in assessment to act on these suggestions. It may encourage dialogue across various scholars studying mathematics education topics, such as mathematics educators, psychologists, psychometricians, policy experts, those working in industry, and others.

As authors, we want those working in the K-12 quantitative content measure space to pay more careful attention to the creation and use of K-12 content measures when it comes to validity evidence and interpretations from measures' scores. We share three key ideas that will connect to this need. First, continuing to report test content evidence and reliability is wholly insufficient and instead, scholars must think holistically about the evidence needed to ground score interpretations in robust validity evidence. Thinking critically at the onset of measure refinement and development about what evidence is needed, and how it might be used, better grounds a validity argument and answers the question: How do we know that the interpretations are appropriately grounded in justifiable validity evidence? Measure refinement and development is not an add-on to a research program and instead, should be treated as its own necessary research agenda. Second, as a field of scholars, we must change our language from describing a "content test is valid" to "score interpretations from a content test have validity," or at a minimum "scores from the content test have validity." Modern standards and expectations of measurement from three major organizations have described this clearly, which requires a different perspective than one used in the last 50 years. As a field of scholars, we can change this through education about modern validity standards and validity arguments for graduate students within STEM contexts and within peer-review contexts for journals and conferences that impact mathematics education scholarship. Third, the current emphasis on formative and classroom assessments of K-12 students' content knowledge requires rethinking what validity evidence is warranted for desired score interpretations and how we collect that evidence. Technology use such as eye-tracking while students complete a content test aimed at formative information provides opportunities to potentially connect validity evidence to response processes. As a field of scholars, we must spend more time than we are currently spending, thinking about the validity evidence associated with the outcome interpretations from formative and classroom assessments used in the classroom. A shared understanding of validity arguments may be stimulated by working across contexts on a shared problem. Thus, we encourage readers to consider the historical problems as well as the suggestions provided in this chapter as opportunities to develop into research agendas, which will ultimately have broad impact across scholars working on mathematics education topics.

References

American Educational Research Association, American Psychological Association, & National Council on Measurement in Education. (2014). *Standards for educational and psychological testing.* Washington, DC: American Educational Research Association.

American Educational Research Association, American Psychological Association, National Council on Measurement in Education, & Joint Committee on Standards for Educational and Psychological Testing (U.S.). (1999). *Standards for educational and psychological testing.* Washington, DC: American Educational Research Association.

Behrens, J. T., Mislevy, R. J., Bauer, M., Williamson, D. M., & Levy, R. (2004). Introduction to evidence centered design and lessons learned from its application in a global e-learning program. *International Journal of Testing, 4*(4), 295–301.

Bennett, R. E. (2011). *Automated scoring of constructed-response literacy and mathematics items.* Retrieved from www.ets.org/s/k12/pdf/k12_commonassess_automated_scoring_math.pdf

Bennett, R. E., & Sebrechts, M. M. (1996). The accuracy of expert-system diagnoses of mathematical problem solutions. *Applied Measurement in Education, 9*(2), 133–150.

Bennett, R. E., Steffen, M., Singley, M. K., Morley, M., & Jacquemin, D. (1997). Evaluating an automatically scorable, open ended response type for measuring mathematical reasoning in computer adaptive tests. *Journal of Educational Measurement, 34*(2), 162–176.

Black, P., & Wiliam, D. (2009). Developing the theory of formative assessment. *Educational Assessment, Evaluation and Accountability (Formerly: Journal of Personnel Evaluation in Education), 21*(1), 5.

Bostic, J. D., Krupa, E. E., & Shih, J. (2019). *Assessment in mathematics education contexts: Theoretical frameworks and new directions.* New York, NY: Routledge.

Carney, M., Smith, E., Hughes, G., Brendefur, J., & Crawford, A. (2016). Influence of proportional number relationships on item accessibility and students' strategies. *Mathematics Education Research Journal, 28*(4), 503–522.

Chan, J. Y.-C., & Mazzocco, M. M. (2017). Competing features influence children's attention to number. *Journal of Experimental Child Psychology, 156,* 62–81.

Clauser, B. E., Kane, M. T., & Swanson, D. B. (2002). Validity issues for performance-based tests scored with computer-automated scoring systems. *Applied Measurement in Education, 15*(4), 413–432. doi:10.1207/S15324818AME1504_05

Common Core State Standards Initiative. (2010). *Common Core State Standards for mathematics.* Washington, DC: Author. www.corestandards.org.

Connor, C. M., Mazzocco, M. M., Kurz, T., Crowe, E. C., Tighe, E. L., Wood, T. S., & Morrison, F. J. (2018). Using assessment to individualize early mathematics instruction. *Journal of School Psychology, 66,* 97–113.

Creswell, J. (2012). *Education research: Planning, conducting, and evaluating quantitative and qualitative research* (4th ed.). Boston, MA: Pearson.

Crocker, L., & Algina, J. (2006). *Introduction to classical and modern test theory.* Mason, OH: Thomson Wadsworth.

Demars, C. E. (2007). Changes in rapid-guessing behavior over a series of assessments. *Educational Assessment, 12*(1), 23–45.

Fennell, F., Kobett, E., & Wray, J. (2017). *The formative 5: Everyday assessment techniques for every math classroom.* Thousand Oaks, CA: Corwin.

Gavin, M. K., Casa, T. M., Adelson, J. L. Firmender, J.M. (2013). The impact of challenging geometry and measurement units on the achievement of grade 2 students. *Journal for Research in Mathematics Education, 44*(3), 478–509.

Gobert, J. D., Sao Pedro, M. A., Raziuddin, J., & Baker, R. S. (2013). From log files to assessment metrics: Measuring students' science inquiry skills using educational data mining. *Journal of the Learning Sciences, 22*(4), 521–563.

Gobert, J. D., Sao Pedro, M. A., Baker, R. S., Toto, E., & Montalvo, O. (2012). Leveraging educational data mining for real-time performance assessment of scientific inquiry skills within microworlds. *Journal of Educational Data Mining, 4*(1), 104–143.

Gorin, J. (2006). *Using alternative data sources to inform item difficulty modelling.* Paper presented at the annual meeting of the National Council on Measurement in Education, San Francisco, CA.

Gorin, J. (2014). Assessment as evidential reasoning. *Teachers College Record, 116*(11), 1–26.

Gunderson, E., Park, D., Maloney, E., Beilock, S., & Levine, S. (2018). Reciprocal relations among motivational frameworks, math anxiety, and math achievement in early elementary school. *Journal of Cognition and Development, 19*(1), 21–46.

Hatch, A. (2002). *Doing qualitative research in education settings.* Albany, NY: State University of New York Press.

Hill, H., & Shih, J. (2009). Examining the quality of statistical mathematics education research. *Journal of Research in Mathematics Education, 40*(3), 241–250.

Ivie, J., Kupzyk, K., & Embretson, S. (2004). *Predicting strategies for solving multiple-choice quantitative reasoning items: An eye tracker study: Final report of cognitive components study.* Princeton, NJ: Educational Testing Service.

Kane, M. T. (2001). Current concerns in validity theory. *Journal of Educational Measurement, 38*, 319–342.

Kane, M. T. (2006). Validation. In R. L. Brennan, National Council on Measurement in Education, & American Council on Education (Eds.), *Educational measurement.* Westport, CT: Praeger Publishers.

Kane, M. T. (2012). All validity is construct validity: Or is it? *Measurement: Interdisciplinary Research and Perspectives, 10*(1–2), 66–70.

Kane, M. T. (2016). Validation strategies: Delineating and validating proposed interpretations and uses of test scores. In S. Lane, M. Raymond, & T. M. Haladyna (Eds.), *Handbook of test development* (2nd ed., pp. 64–80). New York, NY: Routledge.

Kilpatrick, J., Swafford, J., & Findell, B. (2001). *Adding it up: Helping children learn mathematics.* Washington, DC: National Academy Press.

Krupa, E. E. (2011). *An executive summary from the conference "Moving Forward Together: Curriculum & Assessment and the Common Core State Standards for*

Mathematics". Arlington, VA: Consortium for Mathematics and Its Applications (COMAP), Inc.

Krupa, E. E., Webel, C., & McManus, J. (2015). Evaluating the impact of computer-based and traditional learning environments on students' knowledge of algebra. *Problems, Resources, and Issues in Mathematics Undergraduate Studies, 25*(1), 13–30.

Mislevy, R. J. (2016). How developments in psychology and technology challenge validity argumentation. *Journal of Educational Measurement, 53*(3), 265–292. doi:10.1111/jedm.12117

Mislevy, R. J., Behrens, J. T., Dicerbo, K. E., & Levy, R. (2012). Design and discovery in educational assessment: Evidence-centered design, psychometrics, and educational data mining. *Journal of Educational Data Mining, 4*(1), 11–48.

National Council of Teachers of Mathematics. (2009). *Focus in high school mathematics: Reasoning and sense making*. Reston, VA: Author.

National Council of Teachers of Mathematics. (2014). *Principles to actions: Ensuring mathematical success for all*. Reston, VA: Author.

National Council of Teachers of Mathematics. (2018). *Taking action: Implementing effective mathematics teaching practices in grades 9–12*. Reston, VA: Author.

National Research Council. (2001). *Knowing what students know: The science and design of educational assessment*. Washington, DC: National Academies Press.

Post, T. R., & Brennan, M. L. (1976). An experimental study of the effectiveness of a formal versus an informal presentation of a general heuristic process on problem solving in tenth-grade geometry. *Journal for Research in Mathematics Education, 7*(1), 59–64.

Pellegrino, J., DiBello, L., & Goldman, S. (2016). A framework for conceptualizing and evaluating the validity of instructionally relevant assessments. *Educational Psychologist, 51*(1), 59–81.

Quellmalz, E., Timms, M., & Schneider, S. (2009). Assessment of student learning in science simulations and games. *WestEd Report to the National Academies, Downloaded, 12*(3), 9.

Roberge, J. J., & Flexer, B. K. (1983). Cognitive style, operativity, and mathematics achievement. *Journal for Research in Mathematics Education, 14*(4), 344–353.

Sample McMeeking, L., Orsi, R., & Cobb, R. B. (2012). Effects of a teacher professional development program on the mathematics achievement of middle school students. *Journal for Research in Mathematics Education, 43*(2), 159–181.

Sebrechts, M. M., Bennett, R. E., & Rock, D. A. (1991). Agreement between expert-system and human raters' scores on complex constructed-response quantitative items. *Journal of Applied Psychology, 76*(6), 856.

Selke, D., Behr, M., & Voelker, A. (1991). Using data tables to represent and solve multiplicative story problems. *Journal for Research in Mathematics Education, 22*(1), 30–38.

Shepard, L. (2018). Learning progressions as tools for assessment and learning. *Applied Measurement in Education, 31*(2), 165–174.

Shermis, M. D., & Burstein, J. C. (2003). *Automated essay scoring: A cross-disciplinary perspective*. Mahwah, NJ: Lawrence Erlbaum Associates, Inc.

Shumway, R., Wheatley, G., Coburn, T., White, A., Reys, R., & Schoen, H. (1981). Initial effect of calculators in elementary school mathematics. *Journal for Research in Mathematics Education, 12*(2), 119–141.

Shute, V. J. (2011). Stealth assessment in computer-based games to support learning. *Computer Games and Instruction, 55*(2), 503–524.

Szetela, W., & Super, D. (1987). Calculators and instruction in problem solving in grade 7. *Journal for Research in Mathematics Education, 18*(3), 215–229.

Tosto, M. G., Petrill, S. A., Malykh, S., Malki, K., Haworth, C. M. A., Mazzocco, M. M. M., . . . Kovas, Y. (2017). Number sense and mathematics: Which, when and how? *Developmental Psychology 53*(10), 1924–1939. doi: 10.1037/dev0000331

Watts, T., Clements, D., Sarama, J., Wolfe, C., Spitler, M., & Bailey, D. (2017). Does early mathematics intervention change the processes underlying children's learning? *Journal of Research on Educational Effectiveness, 10*(1), 96–115.

Way, W. D., McClarty, K. L., Murphy, D., Keng, L., & Fuhrken, C. (2011). *Through-course common core assessments in the United States: Can summative assessment be formative?* Paper presented at the annual meeting of the American Educational Research Association, New Orleans, LA.

Webel, C., Krupa, E. E., & McManus, J. (2017). The math emporium: Effective for whom, and for what? *International Journal of Research in Undergraduate Mathematics Education, 3*(2), 355–380.

Williams, S., & Leatham, K. (2017). Journal quality in mathematics education. *Journal of Research in Mathematics Education, 48*(4), 369–396.

Wilson, M. (2005). *Constructing measures: An item response modeling approach.* Mahwah, NJ: Lawrence Erlbaum Associates.

Wise, S. L., & Demars, C. E. (2006). An application of item response time: The effort moderated IRT model. *Journal of Educational Measurement, 43*(1), 19–38.

10 Future Directions in the Measurement of Mathematics Teachers' Competencies

Heather Howell, Elizabeth Stone, and Michael Kane

Shifts in the types of assessments that measure mathematics teachers' competencies have implications for validity arguments around those assessments. Messick defined assessment validity as "an integrated evaluative judgment of the degree to which empirical evidence and theoretical rationales support the *adequacy* and *appropriateness* of *inferences* and *actions* based on test scores or other modes of assessment" (Messick, 1989, p. 13, emphasis in original). While certain types of evidence are widely utilized and accepted as part of that support, examining validity of a given assessment's outcomes is less a prescribed application of methods and more a rigorous and thoughtful process of specifying the intended interpretations and uses of the scores and using a validity framework to map evidence to those intended interpretations and uses. Validation requires constant attention to the strength and relevance of the evidence, the strength of the rationale connecting it to the interpretation, and the consideration of evidence to the contrary (Kane, 2006). It is further recommended that test developers clearly document intended use and supported interpretations (American Educational Research Association, American Psychological Association, & National Council on Measurement in Education, 2014). All of this implies that shifts in constructs to be measured, methods of measuring those constructs, and intent of interpretation and use require a re-evaluation of the associated validity arguments to account for those changing factors.

We focus in this chapter on three trends in the field of teaching competency assessment that have implications for how the validity of those assessments is evaluated and supported. These three trends were selected because they are at different stages of maturity, illustrate particular challenges to validity arguments, and have the potential to substantially shift the landscape of teacher assessment. One well-established movement is a shift in focus from capturing only teachers' mathematical content knowledge toward capturing their mathematical knowledge for teaching (MKT). A second and more developing movement is the trend toward measuring practical competencies through performance measures, which follows a widespread shift in teacher education toward practice-based programs of study. A final movement is the push toward formative assessment of teaching competency to improve MKT and teaching practice, which we describe as emergent in that there is a growing consensus around the importance of such assessment but not yet a strong infrastructure in place to support it. We highlight three different types of validity considerations in these trends: shifts in the types of constructs targeted by assessment, considerations around construct relevance and operationalization in new testing formats, and considerations of use and interpretation in a formative (rather than summative) framework.

Throughout the chapter, we deliberately keep our discussion of validity agnostic with respect to specific frameworks and avoid the use of technical language where possible. General approaches to systematically accounting for validity concerns are present in the literature, and we encourage interested readers to explore those frameworks more fully. For example, Kane's (2006) validity argument approach involves laying out an interpretive argument to specify the reasoning that connects observed performances to inferences, and it provides vocabulary for describing the high-level categories of inferences one would expect to see. Schilling and Hill (2007), while taking up the same fundamental set of assumptions and inferences, argued for a differently organized set of high-level categories. The *Standards for Educational and Psychological Testing* (American Educational Research Association et al., 2014) itself might be taken as a framework by which to delineate, for a given assessment, the sources of validity evidence that might be examined. The common element across these approaches is the underlying notion that in order to consider the validity of an assessment's outcomes and interpretations, one must make a careful, rigorous, and clear effort to link the evidence collected to support the assessment with the desired claims. In all cases, this includes attention to evidence of appropriate construct measurement, adequate sampling of items or tasks from the target domain for the intended conclusions, and justifiable interpretations and uses. Some frameworks may be more amenable to particular assessment types, but the larger point is that collection of validity evidence should be systematic, thorough, and transparent. In our discussion, we do not focus on a delineating a complete validity argument but rather call the reader's attention to aspects of these shifts in assessment that are most likely challenging from the perspective of validity.

In order to explore all three trends as deeply as possible within the boundaries of this chapter, we limit our scope to assessments of teachers' mathematics knowledge, mathematical knowledge for teaching, and mathematical teaching competency, although we recognize that there are many teaching competencies that transcend subject matter and are of critical importance to quality teaching (Sykes & Wilson, 2015). We consider challenges, opportunities, or open questions about the ways in which we will need to attend to the validity evidence for such assessments moving forward. Our goal is not to provide a comprehensive overview of teacher assessment or validity; rather, we seek to illustrate the trends using specific example assessments and feedback mechanisms as a springboard to identify validity concerns and to clarify points on which we argue that more thinking is needed.

Trend 1: A Shift to Measuring Mathematical Knowledge for Teaching

We begin with a shift in the field toward the measurement of teachers' mathematical knowledge for teaching (MKT) rather than, or in addition to, their knowledge of mathematics. This shift has been underway for some time, with foundational research dating back decades (e.g., Ball, 1990) and well-established examples of accompanying assessments (e.g., Hill, Schilling, & Ball, 2004). MKT is broadly accepted as a type of knowledge teachers need and that can be assessed in valid and reliable ways. Standards documents (e.g., Association of Mathematics Teacher Educators, 2017; Conference Board of the Mathematical Sciences, 2012) recognize MKT as a critical body of knowledge for teachers to master during their preparation programs. The Praxis teacher licensure series of tests now includes an elementary-level assessment that measures MKT as well as similar knowledge in other subjects (Educational Testing Service, n.d.a). Public policy reflects a need for valid and reliable assessments at scale across all grade levels (National Research Council, 2013).

Construct Definition: What Is MKT?

Construct definition is worth discussing in more detail, as it is complex in ways that must be accounted for in evaluating assessment validity. One complication is that terminology is not universal; definitions used in various studies often share terminology but not meaning (Hill, Sleep, Lewis, & Ball, 2007; Kaarstein, 2014).

In this chapter, we use terminology as defined by Ball, Thames, and Phelps (2008). MKT is defined as mathematical knowledge that a teacher must draw on in doing the recurring work of teaching (Ball et al., 2008), including Pedagogical Content Knowledge (PCK) and subject-matter knowledge, which includes both Common Content Knowledge and Specialized Content Knowledge (SCK). SCK is "'specialized' because it is not needed or used in settings other than mathematics teaching" (Ball et al., 2008, p. 396). For example, familiarity with the standard algorithm for subtraction is common content knowledge, but being able to make sense of the types of (correct and incorrect) non-standard approaches to subtraction that students might take is SCK. It is content knowledge (i.e., not PCK) because making sense of mathematical solutions is fundamentally mathematics; an individual with the appropriate mathematical knowledge could make sense of a non-standard approach and decide whether it is viable without necessarily knowing why a child would think about it that way or whether it is a common approach that students take. It is specialized (to teaching) because it is difficult to imagine a context in which non-teachers would necessarily need to engage in that type of mathematical thinking, but teachers need to be able to do so fluently and frequently in their work.

We note that the majority of assessments that measure MKT focus to a large degree on items that measure mathematics content knowledge, but it is the type of mathematics that is closely tied to the work of teaching. This may be because items that measure (within the MKT construct) something more like pure mathematics tend to have correct answers that can be logically justified on a mathematical basis. Those that measure (within the MKT construct) something more like pedagogy rely more on judgment about best pedagogical practices. Because these judgments are generally more subjective, it can be harder to justify such answers. Perhaps as a result, most of the MKT assessments for which validity evidence has been gathered tend to target knowledge of the first type.

Assessment Frameworks for MKT

More recent assessment development efforts have adopted a systematic approach in which MKT is taken to be "co-defined" by the student-level mathematics and the work of teaching (Selling, Garcia, & Ball, 2016). For example, in designing an assessment item, one might specify a type of work the teacher is engaged in (e.g., analyzing written student work) and an idea in mathematics that students might be learning (e.g., multi-digit subtraction). The item targets the knowledge the teacher holds at the intersection of these two (e.g., the mathematics the teacher needs to know in order to analyze students' written approaches to multidigit subtraction), and the structure is essentially a matrix with cells representing those intersections. It remains the work of the assessment developer to decide, item by item, what piece of mathematical knowledge among those co-defined by the two framework dimensions the item should focus on measuring. A strength of the underlying theory has been that defining MKT by the context in which it is needed maintains the strong connection to teaching practice. A largely untested assumption implicit in this framing is that the two dimensions are adequate to support sampling, both in the sense that they call out the meaningful distinctions for sampling and in the sense that each cell is sufficiently populated with knowledge targets that a reasonable pool of assessment items can be created.

Implications for Validity of MKT Assessments

Assessments of MKT can generally make use of traditional assessment formats and short performance tasks, and the resulting scores can be interpreted in terms of a fairly well-defined MKT domain. Therefore, in such relatively simple situations, the interpretation of scores is correspondingly direct, and the network of inferences and assumptions leading from the observed performance to the score is also relatively simple and direct. Likewise, many of the approaches to collecting validity evidence for such an assessment would use established procedures such as ensuring adequate processes to produce and document defensible correct answers for the assessment tasks, conducting generalizability and reliability studies, and attending to use and interpretation by conducting studies to ensure that both the content of the tests and any cutoff scores are reflective of the competence needed for teachers at particular grade levels and levels of experience.

We discuss here several potential wrinkles relating to construct validity and considerations of construct-irrelevant variance, as well as to the response processes used by test takers. As described previously, the MKT construct is generally defined indirectly as part of assessment frameworks. This means that in examining evidence of whether the sample of performances is representative of the domain, scholars would need to attend to both dimensions—the work of teaching and the content of that teaching. In addition, development of domain models and subsequent expert review would need to attend to the ways in which the work of teaching and the content intersect, as some intersections might contain mathematics of varying importance as components of effective teaching.

The second complicated aspect is construct relevance, which requires that the assessment reflect only what one claims to be measuring. One reason that defining the construct indirectly through its context may be effective is that most assessments of MKT follow what is referred to as a practice-based item design, in which a teaching scenario is described in text for the test taker to respond to. Specification of the work of teaching paired with the student mathematics provides a natural foundation for that teaching scenario, suggesting what the students might be doing as they engage in mathematics and what the teacher will need to do to respond.

While such scenarios are not always lengthy, there is a tendency for the items both to be text-intensive and to require close reading and attention to specific wordings to cue the set of understandings intended by the item designer. This design feature may introduce construct-irrelevant variance. A MKT test that uses difficult or complex language is at risk of measuring how closely the test taker read the text along with the test taker's MKT, because incorrect answers might reflect deficiencies in either. Arguments can be made to soften this concern. First, one can consider whether the reading load is unreasonable for the target population, college-educated adults who are or are preparing to be teachers. In other words, the language may not be too complex even if it is complex. As a general principle, items designed to measure mathematical content knowledge should not include more complex language than is needed.

A third consideration in the use and study of practice-based MKT items is that of student response processes. There is some evidence that there can be multiple cognitive paths leading to a single correct answer in some cases. For example, Lai and Howell (2016) described an assessment item in which the teacher is asked to evaluate written answers given by two students. A key piece of information the test taker must pay attention to is that the problem the students were given did not specify a set of data to be linear, and one student's solution assumes that it is. This piece of information, however, can be noted in different ways by teachers responding to the test item. One test taker might consider it from a mathematical point of view, recognizing the linearity assumption as critical. Another might consider it from a pedagogical point of view, recognizing that this is a common error for students. While it is essential that the test taker notice this detail, how or why this happens may not matter. However, if a correct answer can be reached via quite different paths, one that cues mathematical knowledge and another that cues pedagogical experience, this could pose a validity concern if the inferences to be made assumes one path over the other.

A fourth and related consideration is the measurement of subcomponents of MKT. Many assessments that use practice-based items have been designed with the explicit goal of measuring mathematical knowledge and PCK on separate assessment forms (e.g., COACTIV: Krauss, Baumert, & Blum, 2008; DTAMS: Saderholm, Ronau, Brown, & Collins, 2010). Others have been conceptualized under an assumption that each item assesses one knowledge subdomain in isolation from others, which has been noted as potentially problematic (e.g., Howell, Lai, & Suh, 2017). For assessments that rely on this assumption, items that elicit multiple correct response pathways by drawing on different knowledge components could represent a significant validity challenge, as could items that measure multiple subdomains simultaneously. On the other hand, evidence that test takers draw on different knowledge subdomains of MKT is likely not problematic if assessments are designed to capture MKT as a whole.

We have described this trend as well established because assessments of teacher MKT are rapidly becoming mainstream. To a great extent, these assessments resemble other content tests and do not present unusual validity challenges. However, because MKT is indirectly and contextually defined, special attention to construct validity is necessary, as is attention to the possibility of irrelevant variance introduced by the commonly used practice-based item design. Further, assessments of MKT that depend on measuring distinct subdomains of the construct likely require greater attention to construct validity and validity with respect to response processes, as practice-based item formats may support multiple reasoning paths.

Trend 2: Assessing Teaching Practice

A convergence of factors has begun to make performance assessments attractive and practical in new ways. The most salient is a movement in teacher education toward what is called practice-based teacher education, in which there is an increased focus on the work that teachers need to do (e.g., Association of Mathematics Teacher Educators, 2017; Ball & Forzani, 2009; Grossman, 2018). Another factor is a movement in teacher testing to create assessments that are in some way "closer to teaching" (e.g., Kersting, Givvin, Thompson, Santagata, & Stigler, 2012; Phelps & Howell, 2016; Shaughnessy & Boerst, 2018), where this generally means that the constructs measured clearly map onto the work teachers need to do, and the resulting assessments ask teachers to act out some part of teaching practice. The fidelity to teaching practice may vary. In this section, we explore newer ways of testing constructs such as teaching practice through more authentic proxies, and we describe aspects of these less traditional testing formats such as the fidelity of the proxy and the alignment of scoring to intended interpretations that may result in validity challenges.

Practice-Based Teacher Education and Its Influence on Assessment

The notion of practice-based teacher education is strongly associated with a theory of learning in and from practice that describes how novice teachers can learn from engagement in core practices that make up the work of teaching (Grossman, Compton, Igra, Ronfeldt, Shahan, & Williamson, 2009). Pedagogies that support such learning include representations and approximations of practice. *Representations* are examples of teaching and vary in nature and grain size, ranging from observations of teaching to "brief narrative accounts of a constructed classroom dilemma meant to provoke problem-solving" (Grossman et al., 2009, p. 2065). *Approximations* engage the teacher in rehearsal of some constrained version of teaching, with higher fidelity examples such as supervised student teaching and lower fidelity examples such as "analyzing a written case" (Grossman et al., 2009, p. 2079). This theory has been associated with efforts to refocus programs of teacher preparation around such engagements, in contrast to a more traditional organization around coursework designed to help prospective teachers accumulate relevant knowledge. This idea has widespread support in the field. Ball and Forzani (2009), for example, call for teacher education to "emphasize repeated opportunities for novices to practice carrying out the interactive work of teaching" (p. 500), and scholars have noted that teacher candidates are more likely to be effective when their preparation is directly linked to classroom practice (Association of Mathematics Teacher Educators, 2017; Boyd, Grossman, Lankford, Loeb, & Wyckoff, 2009).

The influence of practice-based teacher education on assessment can be thought of in two interrelated ways. The first influence is a shift in values toward favoring what teachers can *do* more strongly relative to what they *know*. This shift is simultaneously subtle and dramatic. Knowledge and the ability to use that knowledge in practice are interrelated, and most teacher preparations programs, traditionally and currently, have a mixed focus that includes both. But what is new is that teaching practice has not generally been used as the organizing principle for programs of teacher preparation.

The second influence that practice-based teacher education has had on assessment is through the natural extension of these ideas to assessment format. For example, it has long been the case that teachers' professional training included reflecting on representations of practice such as video recordings (Maher, 2008) or instructional cases (Sykes & Bird, 1992). Projects like Kersting et al.'s (2012) work on analysis of classroom video extend that activity to an assessment format, asking test takers to make judgments about representations of practice. The underlying theory is that by eliciting responses to test stimuli that are more immediately representative of classroom teaching, this assessment measures something closer to the ability to teach than a purely knowledge-based measure might.

Performance measures that engage teachers in approximations of practice arguably measure the ability to do the work of teaching even more directly, and a number of programs have developed such assessments. For example, at the University of Michigan, pre-service teachers sit for a competency exam in which they interview an adult pretending to be a student in order to determine what the "student" understands (Shaughnessy & Boerst, 2018). A program is currently in development by the Woodrow Wilson program at the Massachusetts Institute of Technology, with plans to organize around a mastery-based approach in which the result of a series of performance assessments is the basis for advancing (Woodrow Wilson National Fellowship Foundation, n.d.). Performance-based assessment designs are increasingly reaching the mainstream and are being applied outside of university settings. The Educational Testing Service's NOTE examination was designed to include components in which a teacher candidate interacts in real time with five digital student avatars and leads them through a lesson (Educational Testing Service, n.d.b). While the use of such technology is not new (e.g., Dieker, Hynes, Hughes, & Smith, 2008), it represents the first such effort to use the technology in a rigorous assessment format with attention to the types of validity concerns that would be raised should such an assessment be used for higher-stakes decision making.

Implications for the Validity of Outcomes and Interpretations for Performance Assessments of Teaching Practice

New assessment approaches imply, in turn, the need to formulate new validity arguments, and familiar themes may play out in different ways.

We again encounter the threat of construct-irrelevant variance and also consider face validity, the notion that an assessment has to seem relevant and valid to the test taker and other stakeholders (e.g., test score users) or risk a perceived lack of value, possibly resulting in disengagement or distraction on the part of the test taker and a reluctance or refusal to accept the score by other stakeholders. While there has been enthusiasm for pursuing more innovative and interactive testing formats (e.g., with technologically enhanced items, simulations, or game-based environments), there needs to be clear evidence to support their use as replacements of or supplements for traditional tests (e.g., multiple-choice tests). It is important that novel test design, development, and administration not outpace the collection of this evidence.

Assessments of teaching skill that succeed in closely approximating the classroom as a teacher would experience it present an attractive opportunity to measure teaching skills that are deeply important but that do not easily decompose into smaller measurable units. However, approximations of naturalistic phenomena, to the extent that they are successful, of necessity incorporate a great deal of variance. Classrooms are, in truth, messy places in which many things happen simultaneously and in which it can be difficult, even for experienced teachers, to attend to everything at the same time. To the extent that a simulation is intended to measure how the teacher contends with a particular type of situation, this messiness may be construct-relevant. But, to the extent that what is to be measured is some subcomponent of teaching, there is a risk of construct-irrelevant variance. And while we reiterate the need to try to measure exactly and only what is intended to be measured, we recognize that this may be difficult in a situation that is designed to look and feel natural, where removing certain aspects of the teaching being simulated may, in turn, compromise face validity.

We caution against two logical errors that are easy to fall into when considering simulations as approximations of classroom teaching. First, it is easy in this assessment format to confound performance on the task with actual teaching. Certainly, simulations are designed to be proxies for teaching, but the skills involved when engaging in one are not identical to the skills involved when engaging in the other. Real teaching, for example, depends on building relationships with students (and families) and establishing classroom norms over time, neither of which is available in an on-demand simulated teaching assessment.

A related assumption is the notion that usefulness increases with fidelity to teaching. One of the affordances of simulation for assessment purposes is the ability to be deliberate with respect to which parameters resemble real-word phenomena and which do not. For example, in an approximation of teaching, student misbehavior can be left in or filtered out to allow a focus on behavior management skill or a focus on content teaching, respectively. However, the same principles around reducing complexity to the extent possible apply here as they would in any other assessment context. The more detail and complexity we add to an assessment context, the more difficult it tends to be to score performances consistently, and the easier it is for sources of construct-irrelevant variance to be introduced. Also, because high-fidelity performance tasks tend to be long and complicated, the resulting scores may be less reliable than they are in a more standardized format, and reliability is an important concern in high-stakes contexts. The extent to which enhanced realism is useful can vary tremendously. One might argue that if the point is to evaluate whether an individual can teach under real-world conditions, the only thing that matters is whether he or she can teach under real-world conditions, and everything else is (potentially) a poor proxy only valuable in so far as it is predictive. However, for formative uses described in the following section, the value of the assessment might be in disentangling the web of knowledge, skills, and abilities so that a prospective teacher knows where to focus.

With any new type of technology comes the potential to introduce entirely new sources of (potentially construct-irrelevant) variance. Several of the simulation-based assessments cited above depend on human beings who may themselves pretend to be students or control student avatars in a way that conceals their presence. This introduces an additional source of variance (the human being) into the test architecture. This requires test designers to think carefully about what standardization means in such a context and whether the human components of the test can be adequately standardized to support reliable scoring. Mikeska, Howell, and Straub (2017) describe how a performance assessment of teachers' ability to lead discussions pushes our thinking about standardization in other ways. One example they raised is that the longer and more interactive a performance is, the more ways there are that it can play out very differently for different test takers, resulting in situations where two test takers have, by the end of the interaction, experienced two quite different tests.

Depending on what the test taker does and how well he or she does it, what is required as the simulation proceeds may also become relatively easier or harder over time, so a test taker whose performance is weaker out of the gate may encounter more challenges subsequently than does a test taker whose performance is stronger initially. And while steps can be taken to constrain variation in task design, the approximation must ultimately feel sufficiently natural to meet the face validity bar—too much constraint might undermine the assessment task's validity in other ways. This does not imply of necessity that the scores are not comparable. In adaptive testing, while test takers are administered different items, the scores that result are comparable because the item parameters are calibrated to ensure that they will be. Depending on the construct of interest, the opportunity to observe and make judgments about the test taker's performance may not be significantly undermined by less standardization.

The larger point is that it is important that the resulting scores reflect what is valued, and this may require constraints within the assessment. For example, if capable classroom management is the construct of interest, then we might decide that the algorithm underlying the simulation should not reward a test taker who stumbles initially and recovers with a heroic effort at the end over a teacher who maintains the classroom environment consistently and at a high level of quality throughout. Again, design of the assessment and the scoring rubric should be aligned with the desired uses and interpretations of the results.

We would argue that a validity argument for an interactive performance assessment should, along with including aspects in common with other assessment types, attend especially to the nature and extent of the approximation; should provide a justification for why performance is necessary for capturing the needed information about the target skill; and should have a logic that maps out the relevant parameters, the degree to which they are to be approximated with fidelity, and the reasoning for those choices. Validity arguments should address, among other considerations, the potential for construct-irrelevant variance as a result of approximating naturalistic phenomena with adequate fidelity, any unexpected potential sources of irrelevant variance such as variability in the behavior of the human beings enacting the approximation, and the question of whether the approximation can be done with sufficient fidelity given the intended score uses.

Trend 3: Formative Assessment of Teaching Competency

The first two trends focused on construct definition and method of assessment. We now turn to the question of use and discuss an emergent trend toward developing more structured types of formative assessment of teacher knowledge and practice.

Formative assessment of students has been studied more extensively than has formative assessment of teachers, and we borrow relevant examples from the student literature while acknowledging some risks. In the context of a theory of action that explains how information supports change, teachers and students may have different characteristics and benefit from different feedback types, structures, and support. While these distinctions do not preclude the basic analogy between student formative assessment and teacher formative assessment, contextual factors are key in the underlying theory of action. This is one reason we will argue that theorization and study of teacher formative assessment and learning are critically needed. That this effort is important is supported by, for example, a 2013 special issue of *Teachers and Teaching* on "Formative Assessment and Teacher Professional Learning" (volume 19, issue 2).

We focus on formative assessment for in-service, rather than pre-service, teachers. The above distinctions notwithstanding, pre-service teachers are generally enrolled, as students, in university coursework designed to support their learning of teaching practice, and it is relatively straightforward to extend a teacher/student model to teacher educator/pre-service teacher. It is less clear that the model is simple to extend to in-service teachers, which makes it the more interesting case to explore in the context of validity implications.

We begin by providing an overview and a grounding example of student formative assessment. We then explore what teacher formative assessment is and could be, specifying sources of formative feedback and other evidence to support improved MKT and other teacher competencies. Finally, we focus on challenges with respect to validity and efficacy of teacher formative assessment, with special attention to the connection to a sound theory of learning.

A Brief Overview of Formative Assessment

Black and Wiliam (1998) defined formative assessment of students "as encompassing all those activities undertaken by teachers, and/or by their students, which provide information to be used as feedback to modify the teaching and learning activities in which they are engaged" (pp. 7–8). We note three salient points in this definition that will help frame our discussion. The first point is that the definition encompasses many activities which traditionally might be considered instruction rather than assessment. The second point is the implicit assumption that students are assessed and that teacher's role is related to the assessing. Definitions of student formative assessment have since evolved to encompass active roles for students in monitoring their own learning (Heritage, 2008). Similarly, teachers and teaching may be the subject of assessment, and teachers may take an active role in their own learning.

The third point is that the purpose of formative assessment is *modification* via feedback and action. This is one of several important characteristics that differentiates summative from formative assessment. Both are intended to provide information, but summative assessment is often used in the service of evaluation or broader decision making (e.g., year-end level of proficiency, promotion/retention, or graduation), while formative assessment is intended to be instructionally actionable in service of improvement. Penuel and Shepard (2016) also noted that more recent approaches foreground *processes* by which change or improvement may take place, supporting our arguments for explicitly including a theory of action and a learning theory in approaches to validation. Further, we will see that it is critical in both student and teacher contexts to evaluate whether the learning occurs.

Student Formative Assessment Using Learning Progressions

Formative assessment requires an underlying learning theory describing hypothesized change or learning. We will use the example of learning progressions (LPs), often called learning trajectories in mathematics

(Lobato & Walters, 2017), as one example of such a basis for formative assessment which might be useful in the formative assessment of teaching. The National Research Council (NRC) defines LPs as "descriptions of the successively more sophisticated ways of thinking about a topic that can follow one another as children learn about and investigate a topic over a broad span of time" (NRC, 2007, p. 214). We illustrate three key features of formative assessment in the context of an LP-based learning theory in more detail below.

Feedback

Evaluation of where a student is situated on an LP can help to inform next instructional steps (Sztajn, Confrey, Wilson, & Edgington, 2012), in part because placement on an LP provides detailed information about stages of mastery (e.g., that a student may be able to identify a numeric pattern but may not be at the stage to generate a new number that follows the pattern). Placement on an LP not only identifies a point in the trajectory, but by doing so signals a next point in the trajectory as a potential learning target. However, any type of formative assessment information must be accurate enough that unwarranted action is not taken and resources are allocated appropriately. While LPs are based on research, they must be validated empirically by demonstrating that students at a particular level of the LP are able to master tasks aligned to that level, bearing in mind that students may not consistently demonstrate the same level of mastery (see, e.g., Graf and van Rijn, 2016, for an overview of LPs and their empirical validation).

Next Steps

Feedback is only actionable, of course, to the extent that the intended user of the information is able to make use of it. It is not always clear whether teachers employing formative assessment can identify appropriate next steps or know how to implement them (Heritage, Kim, Vendlinski, & Herman, 2009). Providing teacher supports to guide next steps can better prepare teachers to use LPs in informing instruction. For example, the mathematics LPs described in Arieli-Attali, Wylie, and Bauer (2012) are accompanied by "incremental tasks" which are designed to help students to make transitions between levels of the LP. Teachers participating in the study indicated that the tasks helped them to identify specific student misconceptions that might not otherwise have been uncovered. Other empirical studies of teachers' use of LPs in the classroom have shown similar benefits (see, e.g., Clements, Sarama, Spitler, Lange, &Wolfe, 2011; Edgington, 2012; Wilson, Sztajn, Edgington, & Myers, 2015); however, more specific guidance on use may be warranted due to variability in how teachers implement these practices.

Re-evaluation of Learning

The re-evaluation of student learning after targeted instruction provides some evidence of whether instructional interventions linked to formative assessment were successful (see, e.g., Nichols, Meyers, & Burling, 2009). As modification is one of the specific goals of formative assessment, identifying whether student learning took place and can be attributed to the interventions is an important step in the validation process. This may be more complicated to achieve than simply retesting.

What Does (or Might) Teacher Formative Assessment Look Like?

We note again the reasonable expectation that teacher formative assessment may parallel student formative assessment to some degree. For example, the distinction between summative and formative purposes holds for teachers as well as students. An assessment geared at assessing readiness to teach (i.e., licensure assessment) has a very different purpose than one geared to help a teacher learn (e.g., teaching evaluation by a coach). And the sources of information collected in formative assessment of teachers may be similarly broad, ranging from formal testing to any form of communication between teachers and students, colleagues/peers, or administrators that gives the teacher information about the effectiveness of his or her teaching. Similarly, the types of validity challenges that accompany each type of feedback and the evidence that it provides (e.g., is it accurate and actionable?) are just as critical to consider in the formative assessment of teaching.

Interim student assessment data are one type of feedback from students that can be used in evaluating and informing development of teaching practice. An example can be found in the Winsight Assessment System (Wylie, 2017), in which a computer-delivered set of modules for student interim assessment provides feedback mapped to LPs. We note that this is an example of overlap in which student formative assessment and teacher formative assessment are intertwined. The modules provide feedback on student learning designed to help the teacher adjust instruction. However, the information about how students are understanding the intended content of instruction also constitutes a type of formative feedback to the teacher on the teacher's effectiveness. Both interpretations suggest adjustment to instruction and likely include similar challenges in interpretation and implementation, but the theories of action and mechanisms by which student learning and teacher learning are provoked may be quite different. For example, results of the Goertz, Nabors Oláh, and Riggan (2009) study on the use of interim assessments for instructional change indicate that although teachers reported using the information to adjust instruction, their responses also surfaced the beliefs teachers brought to the table about student misconceptions.

We extend the LP example from the student section to illustrate how the use of student LPs in formative assessment of students might also inform the formative assessment of teachers' MKT, as Falk (2012) suggests. In our example, students might respond to tasks aligned to the Equality and Variable LP (Arieli-Attali et al., 2012) providing information to the teacher about the students' conceptions or misconceptions about the "concept of variable." A formative assessment of the students' teacher might focus on the teacher's MKT around the concept of variable as well as the teacher's ability to interpret and act on student understanding, with evidence coming from peer observation of a classroom discussion led by the teacher. These are important and overlapping ideas. The LP placement helps to guide instructional next steps and forms a significant part of the theory of action that leads to student learning. The teacher's MKT is needed in order to interpret and make sense of what the student says, understand why it is consequential, and determine appropriate next instructional steps (see, e.g., Ebby, Sirinides, Supovitz, & Oettinger, 2013; Wilson, Sztajn, Edgington, & Confrey, 2014). Peer feedback to the teacher about how student misconceptions were identified and next instructional steps planned can help to identify gaps in the teacher's MKT and could potentially lead to specific recommendations to support teacher learning. This forms a significant part of the theory of action that leads to teacher professional growth, although we note that the mechanism by which the teacher is assumed to learn remains underspecified.

In this example, we have called attention to how a piece of evidence may simultaneously inform the formative assessment of both students and teachers. Implicit is that the difference in the contexts is in the interpretation and use of that piece of evidence to inform learning. This ties directly in to the notion that the theory of action for change, and the validity evidence required to support it, will differ for student and teacher formative assessment even where the evidence is shared. It can be difficult to disentangle students' learning from teachers' teaching competency. Teaching is, after all, performed in service of student learning, and it is possible to describe a theory of action that accounts for both pieces (see, e.g., Bennett, 2010). However, the serious consideration of teacher formative assessment demands that we clearly specify the theory of action for change in teaching competency.

Implications for Validity of Outcomes from Teacher Formative Assessment

Bennett (2011) argued that a formative assessment's theory of action, which relates claims through a logic model to outcomes for stakeholders, should include both a validity argument and an efficacy argument. The validity argument focuses on the claims about interpretations that are made, and the efficacy argument focuses on claims about how the interpretations can justifiably be used to adjust instruction and in which ways. Nichols et al. (2009) presented a validity framework that includes an "assessment" phase during which student knowledge is measured, an "instructional" phase during which teachers consider and execute next instructional steps, and a "summative" phase that re-evaluates student knowledge to determine whether the feedback and instruction led to learning. Along with the components specified by Bennett (2011), this framework may help guide the evaluation of validity evidence for teacher formative assessment. Again, the aspects that need to be considered in deciding whether the interpretations and uses of a particular formative assessment process are supported depend on the accuracy of feedback, existence of clearly defined next instructional steps to guide change based on this feedback, and opportunity to re-evaluate whether learning has occurred through these steps. The efficacy claim requires evidence that the specific instructional intervention led to change and must additionally account for whether the observed change could have occurred due to other reasons, as this would reduce the confidence in the specific instructional steps as the mechanisms of change (e.g., Bennett, 2011). Using the example of classroom-based LP usage, a validity argument claim could be that peer observation provides an accurate assessment of whether a teacher is proficient in recognizing and addressing student misconceptions about a particular math concept. A claim from the efficacy argument might be that through a specified professional development activity the teacher's proficiency in this area will improve. The effects of the formative assessment of the teacher's performance could be evaluated by re-observing the teacher in a similar context later. One point our example illustrates is that a student LP is available to guide instruction toward the goal of modifying student learning, while an LP for teacher knowledge or practice is not. We emphasize this point because it is representative of the state of the field and is a place where significant progress could be made to support teacher professional learning (Silverman & Thompson, 2008).

Other areas of attention are also supported by the recommendations of Bennett (2010) and Penuel and Shepard (2016). First, in-service teacher formative assessment feedback often comes from sources other than formal testing. It is just as important to demonstrate that those sources provide dependable evidence as it would be to show that a test's scores are reliable, but this may be challenging. For example, the use of student surveys calls into question whether students felt anonymous enough to respond honestly (see, e.g., Shieh & Cefai, 2017), as the accuracy of answers that are not anonymous can be dubious. Student evaluations may also be biased (positively or negatively) by context and motivation. Peers and administrators should be adequately trained to provide actionable feedback, and the process should be monitored to identify any rater effects or unwarranted variability in judgments. Self-study (e.g., Lampert, 2003) requires both that the teacher correctly identifies construct-aligned evidence of student understanding and that the teacher can interpret those results appropriately to understand how his or her teaching practice has influenced these student outcomes. Teacher learning communities, such as those described by Wylie, Lyon, and Goe (2009), can use peer feedback between teachers to provide mutual support and enhance professional growth, but are most effective when there is formalization of actionable goals and accountability, such as a negotiated learning contract (Verberg, Tigelaar, and Verloop, 2013).

Second, even if the feedback provided suggests a need for or encourages learning, it is not always clear in these contexts what the next learning steps should be or whether the learning opportunities exist. This may be one of the most critical aspects of formative assessment of teacher competency to address, and attention to formalizing theories of action may help to make such gaps more readily apparent.

Third, we argue that there should be more attention paid to evaluating whether learning has occurred after the instruction or intervention was applied, and whether it was, in fact, that action that resulted in the learning. This can be challenging even in the student context, and given the many forms in which feedback reaches teachers, it may be even more difficult to systematically determine whether teaching competency has increased. Because feedback typically originates from more than one source at a time, it is critical to be able to parse the effects of the specific learning mechanism(s) on which validity claims are based. This leads us to argue that developing frameworks such as LPs focused on specific learning goals *for teachers* could be helpful in identifying teacher knowledge in particular areas, providing routes for growth, and specifying what changes in learning we would expect to see based on formative assessment.

A more coherent and complete system of quality feedback, theory-aligned learning steps, and appropriate re-evaluation of specific constructs will be key to developing a robust and effective structure for teacher formative assessment. As formative assessment of teachers continues to emerge as a direction for structured development, these validity infrastructures will need to be integrated into the assessment design to support their use.

Conclusions

In this chapter, we have discussed three trends in mathematics teaching competency assessment, which are in various stages of maturity and which call for attention to validity with respect to the contexts and purposes for which they are being developed. One reason we have used examples of trends at different stages of maturity is to exemplify for the reader the types of validity study that are likely to need attention, and the ways in which that agenda looks different in each area. We frame our reiteration of the three trends as questions, because while our discussion in this chapter is meant to provide background with respect to these issues, our main goal is to inspire more research and collaboration across the fields involved in assessing teaching competency.

The first trend describes a shift in measurement target: What knowledge and skills are being assessed with respect to teaching competency? Specifically, we described a shift from measuring mathematical content knowledge exclusively to measuring aspects of mathematical knowledge for teaching. While this trend is the most mature, there has been less attention than might be expected on a number of points. These include the potential variability in assessment validity across differing populations and linking MKT to student outcomes or other measures of teaching quality, which has currently been done to a limited extent and unevenly across grade levels. In other words, even in this relatively mature line of work there remains careful and systematic work to be conducted to address fundamental questions about construct definition and assessment validity. The mainstreaming of these assessments for use in teacher licensure also opens up an opportunity to study the intended and unintended consequences of MKT assessment use, as recommended by the American Educational Research Association et al. (2014).

The second trend involves shifts in measurement methodology: How can we assess the expanded set of competencies under consideration? While teaching competency tests have traditionally had short-answer formats and have focused on knowledge, measuring teacher practice suggests moving to forms of assessment such as situational judgment tests or interactive simulations. We have described this work as developing, as the measures themselves are still relatively new. These more novel testing formats require validity evidence that goes beyond that needed for traditional formats. We argue that for work at this stage, the construction of validity arguments to account for unique features of the assessment type, such as the nature and extent of the approximation of practice, are critical. There are new potential sources of construct-irrelevant variance, it may be difficult to ascertain construct coverage, and adequate levels of reliability are harder to achieve for performance tests and simulations. Equally important are empirical studies carefully constructed to provide validity evidence. Such studies are relatively few to date, perhaps because of the costs associated with technologies such assessments often depend on.

The third trend involves a change in perspective: How can we develop a more structured formative assessment paradigm for teaching competency? This trend potentially overlaps both the first and second, in that it is possible to measure both MKT and teaching practice formatively, but the focus in formative assessment is on providing teachers with actionable feedback that can provoke professional growth. We reiterate here that the scholarship around teacher formative assessment is less developed than that around student formative assessment, and potentially more complicated, as the teacher is most often responsible for self-instruction or remediation. We envision the development of more complete formative assessment systems that include an evaluation phase, an instruction phase, and a re-evaluation phase and are centered on learning goals based on cognitive maps or LPs like those available for student learning. Validity work with these systems needs to account not just for the accuracy of the inferences that can be made during the evaluation and for the theory of action by which it is hypothesized that the feedback will lead to change during instruction, but also for evidence of whether these changes occur. We have described this trend as more emergent and have given it greater attention in the chapter because there is more work to do to build theories and measures with validity implications in mind and to ensure that efficacy does not fall by the wayside.

That the need for a focus on validity surrounding these trends has been recognized is clear from the literature that we have included in this chapter. The shifts we have described in which teacher competencies are being measured, how, and to what end are all deeply important and positive movements, as teachers are the backbone of our educational system and the minute-to-minute conduits for the learning that our students engage in. Our intention was to share this work with a broader audience in order to promote a larger understanding, from our perspective, of what changes are in progress or in store for teaching competency assessment, and to encourage a rigorous focus on validity. While we welcome the benefits these changes bring, we must anticipate and meet the accompanying challenges head on.

References

American Educational Research Association, American Psychological Association, & National Council on Measurement in Education. (2014). *Standards for educational and psychological testing.* Washington, DC: American Educational Research Association.

Arieli-Attali, M., Wylie, E. C., & Bauer, M. I. (2012, April). *The use of three learning progressions in supporting formative assessment in middle school mathematics.* Paper presented at the annual meeting of the American Educational Research Association, Vancouver, Canada.

Association of Mathematics Teacher Educators. (2017). *Standards for preparing teachers of mathematics.* Retrieved from https://amte.net/sites/default/files/SPTM.pdf

Ball, D. L. (1990). The mathematical understandings that prospective teachers bring to teacher education. *The Elementary School Journal, 90*(4), 449–466. doi:10.1086/461626

Ball, D. L., & Forzani, F. M. (2009). The work of teaching and the challenge for teacher education. *Journal of Teacher Education, 60*(5), 497–511.

Ball, D. L., Thames, M. H., & Phelps, G. (2008). Content knowledge for teaching: What makes it special? *Journal of Teacher Education, 59*(5), 389–407. doi:10.1177/0022487108324554

Bennett, R. E. (2010). Cognitively based assessment of, for, and as learning (CBAL): A preliminary theory of action for summative and formative assessment. *Measurement, 8*(2–3), 70–91.

Bennett, R. E. (2011). Formative assessment: A critical review. *Assessment in Education: Principles, Policy & Practice, 18*(1), 5–25.

Black, P., & Wiliam, D. (1998). Assessment and classroom learning. *Assessment in Education: Principles, Policy & Practice, 5*(1), 7–74.

Boyd, D. J., Grossman, P. L., Lankford, H., Loeb, S., & Wyckoff, J. (2009). Teacher preparation and student achievement. *Educational Evaluation and Policy Analysis, 31*(4), 416–440.

Clements, D. H., Sarama, J., Spitler, M. E., Lange, A. A., & Wolfe, C. B. (2011). Mathematics learned by young children in an intervention based on learning trajectories: A large-scale cluster randomized trial. *Journal for Research in Mathematics Education, 42*(2), 127–166.

Conference Board of the Mathematical Sciences. (2012). *The mathematical education of teachers II.* Retrieved from www.cbmsweb.org/archive/MET2/met2.pdf

Dieker, L., Hynes, M., Hughes, C., & Smith, E. (2008). Implications of mixed reality and simulation technologies on special education and teacher preparation. *Focus on Exceptional Children, 40*(5), 1–20.

Ebby, C. B., Sirinides, P., Supovitz, J., & Oettinger, A. (2013). *TASK technical report.* Philadelphia: Consortium for Policy Research in Education.

Edgington, C. P. (2012). *Teachers' uses of a learning trajectory to support attention to students' mathematical thinking.* Raleigh, NC: North Carolina State University.

Educational Testing Service (n.d. a). *Praxis® elementary education: Content Knowledge for Teaching (CKT) tests.* Retrieved from www.ets.org/praxis/about/ckt/

Educational Testing Service (n.d. b) *National observational teaching exam: What is the note assessment?* Retrieved from www.ets.org/note/what

Falk, A. (2012). Teachers learning from professional development in elementary science: Reciprocal relations between formative assessment and pedagogical content knowledge. *Science Education, 96*(2), 265–290.

Goertz, M. E., Nabors Oláh, L., & Riggan, M. (2009). *Can interim assessments be used for instructional change.* CPRE Policy Briefs. Retrieved from http://repository.upenn.edu/cpre_policybriefs/39

Graf, E. A., & van Rijn, P. W. (2016). Learning progressions as a guide for design: Recommendations based on observations from a mathematics assessment. In S. Lane, M. R. Raymond, & T. M. Haladyna (Eds.), *Handbook of test development* (2nd ed., pp. 165–189). New York, NY: Routledge.

Grossman, P. (Ed.). (2018). *Teaching core practices in teacher education.* Cambridge, MA: Harvard Education Press.

Grossman, P., Compton, C., Igra, D., Ronfeldt, M., Shahan, E., & Williamson, P. W. (2009). Teaching practice: A cross-professional perspective. *Teachers College Record, 111*(9), 2055–2100.

Heritage, M. (2008). *Learning progressions: Supporting instruction and formative assessment.* Retrieved from www.k12.wa.us/assessment/ClassroomAssessment Integration/pubdocs/FASTLearningProgressions.pdf

Heritage, M., Kim, J., Vendlinski, T., & Herman, J. (2009). From evidence to action: A seamless process in formative assessment? *Educational Measurement: Issues and Practice, 28*(3), 24–31.

Hill, H. C., Schilling, S., & Ball, D. L. (2004). Developing measures of teachers' mathematics knowledge for teaching. *The Elementary School Journal, 105*(1), 11–30. doi:10.1086/428763

Hill, H. C., Sleep, L., Lewis, J., & Ball, D. (2007). Assessing teachers' mathematical knowledge. In F. Lester (Ed.), *Second handbook of research on mathematics teaching and learning* (pp. 111–156). Charlotte, NC: Information Age Publishing.

Howell, H., Lai, Y., & Suh, H. (2017). Questioning assumptions about the measurability of subdomains of mathematical knowledge for teaching. In A. Weinberg, C. Rasmussen, J. Rabin, M. Wawro, & S. Brown (Eds.), *Proceedings of the 20th annual conference on research in undergraduate mathematics education* (pp. 413–427). San Diego, CA. ISSN 2474-9346.

Kaarstein, H. (2014). A comparison of three frameworks for measuring knowledge for teaching mathematics. *Nordic Studies in Mathematics Education, 19*(1), 23–52.

Kane, M. (2006). Validation. In R. L. Brennan (Ed.), *Educational measurement* (4th ed., pp. 17–64). Westport, CT: American Council on Education and Praeger.

Kersting, N. B., Givvin, K. B., Thompson, B. J., Santagata, R., & Stigler, J. W. (2012). Measuring usable knowledge: Teachers' analyses of mathematics classroom videos predict teaching quality and student learning. *American Educational Research Journal, 49*(3), 568–589. doi:10.3102/0002831212437853

Krauss, S., Baumert, J., & Blum, W. (2008). Secondary mathematics teachers' pedagogical content knowledge and content knowledge: Validation of the COACTIV constructs. *ZDM, 40*(5), 873–892. doi:10.1007/s11858-008-0141-9

Lai, Y., & Howell, H. (2016, October). Conventional courses are not enough for future high school teachers [Blog post]. Retrieved from http://blogs.ams.org/matheducation/2016/10/03/conventional-courses-are-not-enough-for-future-high-school-teachers/

Lampert, M. (2003). *Teaching problems and the problems of teaching*. New Haven, CT: Yale University Press.

Lobato, J., & Walters, C. D. (2017). A taxonomy of approaches to learning trajectories and progressions. In J. Cai (Ed.), *Compendium for research in mathematics education* (pp. 74–101). Reston, VA: National Council of Teachers of Mathematics.

Maher, C. A. (2008). Video recordings as pedagogical tools in mathematics teacher education. *International Handbook of Mathematics Teacher Education, 2*, 65–83.

Messick, S. (1989). Validity. In R. L. Linn (Ed.), *Educational measurement* (3rd ed., pp. 13–103.) New York, NY: American Council on Education and Macmillan.

Mikeska, J., Howell, H., & Straub, C. (2017). Developing elementary teachers' ability to facilitate discussions in science and mathematics via simulated classroom environments. In T. Bousfield, L. Lieker, C. Highes, & M. Hynes (Eds.), *Proceedings of the fifth annual TeachLive Conference* (pp. 65–76). Orlando, FL: TeachLivE at University of Central Florida.

National Research Council. (2007). *Taking science to school: Learning and teaching science in grades K-8*. Washington, DC: National Academies Press.

National Research Council. (2013). *Monitoring progress toward successful K-12 STEM education: A nation advancing?* Washington, DC: National Academies Press.

Nichols, P. D., Meyers, J. L., & Burling, K. S. (2009). A framework for evaluating and planning assessments intended to improve student achievement. *Educational Measurement: Issues and Practice, 28*(3), 14–23.

Penuel, W. R., & Shepard, L. A. (2016). Assessment and teaching. In D. Gitomer & C. Bell (Eds.), *Handbook of research on teaching* (pp. 787–850). Washington, DC: American Educational Research Association.

Phelps, G., & Howell, H. (2016). Assessing mathematical knowledge for teaching: The role of teaching context. *The Mathematics Enthusiast, 13*(1), 52–70.

Saderholm, J., Ronau, R., Brown, E. T., & Collins, G. (2010). Validation of the diagnostic teacher assessment of mathematics and science (DTAMS) instrument. *School Science and Mathematics, 110*(4), 180–192.

Schilling, S. G., & Hill, H. C. (2007). Assessing measures of mathematical knowledge for teaching: A validity argument approach. *Measurement, 5*(2–3), 70–80.

Selling, S. K., Garcia, N., & Ball, D. L. (2016). What does it take to develop assessments of mathematical knowledge for teaching?: Unpacking the mathematical work of teaching. *The Mathematics Enthusiast, 13*(1), 35–51.

Shaughnessy, M., & Boerst, T. (2018). Designing simulations to learn about pre-service teachers' capabilities with eliciting and interpreting student thinking. In G. Stylianides & K. Hino (Eds.), *Research advances in the mathematical education of pre-service elementary teachers* (pp. 125–140). Cham: Springer.

Shieh, J. J., & Cefai, C. (2017). Assessment of learning and teaching in higher education: A case analysis of a university in the south of Europe. *Malta Review of Educational Research, 11*(1), 29–47.

Silverman, J., & Thompson, P. W. (2008). Toward a framework for the development of mathematical knowledge for teaching. *Journal of Mathematics Teacher Education, 11*, 499–511.

Sykes, G., & Bird, T. (1992). Teacher education and the case idea. *Review of Research in Education, 18*, 457–521.

Sykes, G., & Wilson, S. (2015). *How teachers teach: Mapping the terrain of practice*. Princeton, NJ: Educational Testing Service. Retrieved from https://files.eric.ed.gov/fulltext/ED570633.pdf

Sztajn, P., Confrey, J., Wilson, P. H., & Edgington, C. (2012). Learning trajectory based instruction: Toward a theory of teaching. *Educational Researcher, 41*(5), 147–156.

Verberg, C. P., Tigelaar, D. E., & Verloop, N. (2013). Teacher learning through participation in a negotiated assessment procedure. *Teachers and Teaching, 19*(2), 172–187.

Wilson, P. H., Sztajn, P., Edgington, C., & Confrey, J. (2014). Teachers' use of their mathematical knowledge for teaching in learning a mathematics learning trajectory. *Journal of Mathematics Teacher Education, 17*(2), 149–175.

Wilson, P. H., Sztajn, P., Edgington, C., & Myers, M. (2015). Teachers' uses of a learning trajectory in student-centered instructional practices. *Journal of Teacher Education, 66*(3), 227–244.

Woodrow Wilson National Fellowship Foundation. (n.d.). *Woodrow Wilson academy of teaching and learning: About the WW academy*. Retrieved from https://woodrowacademy.org/about/

252   *Heather Howell et al.*

Wylie, E. C. (2017). Winsight™ assessment system: Preliminary theory of action. *ETS Research Report Series (ETS RR-17–26)*. Retrieved from http://online library.wiley.com/doi/10.1002/ets2.12155/full

Wylie, E. C., Lyon, C. J., & Goe, L. (2009). *Teacher professional development focused on formative assessment: Changing teachers, changing schools (ETS RR-09–10)*. Retrieved from https://onlinelibrary.wiley.com/doi/pdf/10.1002/j.2333-8504.2009.tb02167.x

Index

Note: Page numbers in bold indicate tables and page numbers in italic indicate figures on the corresponding pages.